Science in the Forest, Science in the Past

T0256279

Science in the Forest, Science in the Past: Further Interdisciplinary Explorations comprises papers from the second of two workshops involving a group of scholars united in the conviction that the great diversity of knowledge claims and practices for which we have evidence must be taken seriously in their own terms rather than by the yardstick of Western modernity.

Bringing to bear social anthropology, history and philosophy of science, computer science, classics and sinology among other fields, they argue that the use of such dismissive labels as 'magic', 'superstition' and the 'irrational' masks rather than solves the problem and reject counsels of despair which assume or argue that radically alien beliefs are strictly unintelligible to outsiders and can be understood only from within the system in question. At the same time, they accept that how to proceed to a better understanding of the data in question poses a formidable challenge. Key problems identified in the inaugural workshop, whose proceedings were published in *HAU: Journal of Ethnographic Theory* (2019) and in HAU Books (2020), provided the basis for asking how obvious pitfalls might be avoided and a new or revised framework within which to pursue these problems proposed.

The chapters in this book were originally published in *Interdisciplinary Science Reviews*.

Willard McCarty works on relations among the interpretative human and computational sciences. He is Editor of *Interdisciplinary Science Reviews* and founding convenor of the online seminar for digital humanities, *Humanist* (1987–). His current project is a study of what can be done with artificial intelligence to improve curiosity's well-being.

Geoffrey E. R. Lloyd is based at the Needham Research Institute in Cambridge. The most recent of his many general cross-cultural studies of human cognitive experience is *Intelligence and Intelligibility* (2020).

Aparecida Vilaça is a social anthropologist from Museu Nacional, Universidade Federal do Rio de Janeiro. She has been working with the Wari', an indigenous Amazonian people, for the last thirty years. Her main research themes are cannibalism, warfare, conversion to Christianity, schooling and mathematical knowledge.

Science in the Forest, Science in the Past

Further Interdisciplinary Explorations

Edited by
**Willard McCarty, Geoffrey E. R. Lloyd
and Aparecida Vilaça**

Routledge
Taylor & Francis Group

LONDON AND NEW YORK

First published 2022
by Routledge
4 Park Square, Milton Park, Abingdon, Oxon OX14 4RN

and by Routledge
605 Third Avenue, New York, NY 10158

Routledge is an imprint of the Taylor & Francis Group, an informa business

© 2022 Institute of Materials, Minerals and Mining

British Library Cataloguing in Publication Data
A catalogue record for this book is available from the British Library

ISBN: 978-1-032-15071-0 (hbk)
ISBN: 978-1-032-15072-7 (pbk)
ISBN: 978-1-003-24238-3 (ebk)

DOI: 10.4324/9781003242383

Typeset in Minion Pro
by Newgen Publishing UK

Publisher's Note
The publisher accepts responsibility for any inconsistencies that may have arisen during the conversion of this book from journal articles to book chapters, namely the inclusion of journal terminology.

Disclaimer
Every effort has been made to contact copyright holders for their permission to reprint material in this book. The publishers would be grateful to hear from any copyright holder who is not here acknowledged and will undertake to rectify any errors or omissions in future editions of this book.

Contents

Citation Information

The chapters in this book were originally published in *Interdisciplinary Science Reviews*, volume 46, issue 3 (2021). When citing this material, please use the original page numbering for each article, as follows:

Preface
    Willard McCarty
    Interdisciplinary Science Reviews, volume 46, issue 3 (2021), pp. 213–214

Chapter 1
    Introduction
    G. E. R. Lloyd
    Interdisciplinary Science Reviews, volume 46, issue 3 (2021), pp. 215–224

Chapter 2
    Philosophical engagements with distant sciences
    Nicholas Jardine
    Interdisciplinary Science Reviews, volume 46, issue 3 (2021), pp. 225–240

Chapter 3
    Mongolian map-making as practice
    Caroline Humphrey
    Interdisciplinary Science Reviews, volume 46, issue 3 (2021), pp. 241–266

Chapter 4
    Star canoes, voyaging worlds
    Anne Salmond
    Interdisciplinary Science Reviews, volume 46, issue 3 (2021), pp. 267–285

Chapter 5
    Counting generation(s)
    Marilyn Strathern
    Interdisciplinary Science Reviews, volume 46, issue 3 (2021), pp. 286–303

For any permission-related enquiries please visit:
www.tandfonline.com/page/help/permissions

Notes on Contributors

Alan F. Blackwell is Professor of Interdisciplinary Design at the University of Cambridge with degrees in engineering, computer science and psychology.

Addisu Damena is Lecturer in the faculty of computing, Bahir Dar Institute of Technology, Bahir Dar University at Bahir Dar, Ethiopia.

Stephen Hugh-Jones is Emeritus Research Associate of the Cambridge University Department of Social Anthropology, where he also previously taught.

Caroline Humphrey is a social anthropologist who has worked in Russia, Mongolia, China (Inner Mongolia and Heilongjiang), India, Nepal and Ukraine.

Nicholas Jardine is Emeritus Professor of History and Philosophy of the Sciences at the University of Cambridge.

Geoffrey E. R. Lloyd is Emeritus Professor of Ancient Philosophy and Science at the University of Cambridge and Senior Scholar in Residence at the Needham Research Institute.

Willard McCarty is Emeritus Professor of Humanities Computing, King's College London, Editor of *Interdisciplinary Science Reviews* (2008–) and of *Humanist* (1987–), an online seminar for computing in the human sciences.

Francesca Rochberg is Catherine and William L. Magistretti Distinguished Professor of Near Eastern Studies in the Department of Middle Eastern Languages and Cultures and the Office for the History of Science and Technology at the University of California, Berkeley.

Anne Salmond is Distinguished Professor of Maori Studies and Anthropology at the University of Auckland. She has written a number of award-winning books.

Marilyn Strathern is Emeritus Professor of Social Anthropology at the University of Cambridge and has written on gender relations, kinship, social theorizing and Melanesian thought.

Tesfa Tegegne is Assistant Professor of Computer Science at faculty of computing, Bahir Dar Institute of Technology, Bahir Dar University.

Aparecida Vilaça is an anthropologist and professor at the Graduate Program in Social Anthropology at the National Museum of UFRJ.

Preface

The essays in this issue of *ISR* were originally given as papers in the second of the workshop series *Science in the Forest, Science in the Past*, held at the Needham Research Institute, Cambridge. The series set out to explore and learn from diverse ways of knowing across cultures, disciplines and historical periods. The idea for it arose from discussions in 2016 between G. E. R. Lloyd and a group of Brazilian anthropologists, including Aparecida Vilaça (Museu Nacional, Rio de Janeiro), co-organizer of SFSP II. Essays from SFSP I, held in June 2017, were published in a special issue of *HAU: Journal of Anthropological Theory* 9.1 (2019), and then late last year in a HAU Books volume, *Science in the Forest, Science in the Past*, ed. Lloyd and Vilaça.

The second event followed in June 2019. It was organized by Lloyd, Vilaça and myself, with the generous financial support of the Chiang Ching-kuo Foundation for International Scholarly Exchange (Taipei), and at Cambridge, by the Faculty of Classics and Departments of History and Philosophy of Science and Social Anthropology, to whom we express our deepest appreciation. We also want to thank the Museu Nacional and the Conselho Nacional de Desenvolvimento Cien-tífico e Tecnológico (CNPq, Brazil) for their support. We are most grateful to the Needham and its Director, Mei Jianjun, and the staff, especially Susan Bennett, who helped in ways too numerous to count.

Participants were asked to share their drafts with those from whom they would most like to receive commentary. Hence we are indebted to Bob Amsler (Texas at Austin); João Biehl (Princeton); Keith Breckenridge (Witwatersrand); Lorraine Daston (Max-Planck-Institut für Wissenschaftsgeschichte, Berlin); Carlos Fausto (Museu Nacional); Marina Frasca-Spada (Cambridge); Brad Inwood (Yale); Boris Jardine (Cambridge); Christos Lynteris (St Andrews); Jan Christoph Meister (Hamburg); Knut Christian Myhre (Oslo); Sayana Namsaraeva (Cambridge); Anthony Pickles (Cambridge); Tim Smithers; Richard Staley (Cambridge); Anne-Christine Taylor (CNRS); Baasanjav Terbish (Cambridge); Manfred Thaller (Köln); Thomas White (Cambridge). Partici-pants and attendees also served as commentators.

Comments and questions from attendees were invaluable. Hence our great thanks go to them: Julia Cassaniti (Cambridge); Karine Chemla (Paris-Diderot); Dong Qiaosheng (Cambridge); Marina Frasca-Spada (Cambridge); Fu Daiwie (Yang-Ming); Inanna Hamati-Atiyah (Cambridge); Evelyn Fox Keller (Massachu-setts Institute of Technology); Philippa Lang; Reviel Netz (Stanford); Anthony Pickles (Cambridge); Anastasia Piliavsky (King's College London); Simon Schaffer (Cambridge); Henry Shevlin (Cambridge); Shih Ching-fei (Taiwan); Richard Staley (Cambridge); Liba Taub (Cambridge); Federico Valenti (Cambridge); Wu Huiyi (Cambridge); Zhao Jingyi (Cambridge).

There are a number of reasons to be personally grateful for the experience of scho-larly academic life that *Science in the Forest, Science in the Past* has afforded on these occasions. All too often it is something only dreamt of. But when such experiences happen, they show that all the rest, however unintentionally or accidentally so, has been worth the candle. My heartfelt thanks go to everyone involved.

Willard McCarty

Introduction

Geoffrey E. R. Lloyd

In June 2019, a group of scholars from a variety of disciplines, including social anthropology, history and philosophy of science, computer science, classics and sinology, gathered in Cambridge for the second three-day workshop that we entitled somewhat provocatively Science in the Forest, Science in the Past. We were of course very well aware that in English the term 'science' is often appropriated by what we term the natural sciences, although no such restriction applies to the French 'science' nor to the German 'Wissenschaft', for instance. But we were united in our readiness to take seriously the great diversity of knowledge claims and practices for which we have evidence from ethnography and from history, and also by our dissatisfaction with many superficial attempts at what passed as comparatism, many of them flawed by a basic assumption that Western modernity somehow had to provide the yardstick by which everyone else's efforts are to be judged. The use of such dismissive labels as 'magic', 'superstition' and the 'irrational' simply masks the problem and does not solve it. At the same time, we rejected counsels of despair that assumed or argued that radically alien beliefs are strictly unintelligible to outsiders and can only be understood from within the system in question.

But while we all agreed on the unsatisfactoriness of such swift dismissals, we accepted that quite how to proceed to a better understanding of the data in question poses a formidable challenge. In the first workshop in the series, the proceedings of which were published in a Special Issue of HAU in the Spring of 2019 and then as a HAU book in 2020, we identified a number of key problems and began to explore how it might be possible to avoid obvious pitfalls in their resolution and to suggest a new or revised framework within which to pursue the problems.

One major issue that preoccupied our earlier discussions was that associated with the so-called ontological turn, where it has been suggested that there is an important sense in which we are dealing not with alternative explanations of a single world, but with different worlds, although quite what a commitment to such entails has often been opaque. The answers we give to that question have important repercussions on the extent to which any mutual understanding across worlds is possible. While the multiple world hypothesis had the merit of challenging the hegemonic Western assumption that the world is as Western science says it is (or will say it is at some time in the future), it left in its wake the problem of how any communication or understanding between worlds is possible. We can agree perhaps that there is no common measure between them – they are 'incommensurable' in that sense – and we certainly need to appreciate in any event that no entirely theory-free account is ever possible. But it is one thing to admit that no understanding, no translation, is ever perfect, quite another to conclude that strictly speaking nothing can be understood, nothing communicated, that there is no common ground and no comparison is possible. The latter, extreme, position tends to ignore the enormous variety in the contexts in which mundane communication is attempted and appears to enjoy a modest success according to pragmatic criteria. In practice, the invocation of multiple worlds obscures the issues when it focusses too exclusively on the supposed ontological or philosophical presuppositions believed to be in play. At the same time to insist on our all inhabiting the same one world often went with a further assumption that there is in principle just the one true account of it to be had – which takes us straight back to the problems of cross-cultural understanding and of genuinely even-handed interpretation with which we started.

Similar dissatisfaction began to be registered, in our earlier discussions, concerning other generalising hypotheses, such as the idea that there is some Great Divide separating populations or that the key to explaining diversity in knowledge claims and practices is to be found by pursuing the impact of assumed linguistic, ecological or cultural determinants. Under the first head, Goody (1977) made great play with the effect of developments in the technology of communication, especially the advance of literacy. But while such factors should, for sure, be taken into consideration, the effect of much detailed ethnography and ancient history is to undermine any suggestion that there are fundamental differences in the cognitive capacities of different groups of human beings. The notion that what Goody called 'ruminative reflection' is only possible in a literate society flies in the face of the ample evidence for both criticism and scepticism in predominantly oral groups. Their ability to survive and flourish in environments where members of so-called advanced cultures would soon perish has often been pointed out and should never be underestimated.

A similar set of problems renders the old dichotomy between realism and relativism more of a hindrance than a help, at least when it is presented as a binary choice, either one or the other. We are all relativists in the sense that we are all constantly making comparative judgements, both explicitly and implicitly, and of course at the minimum we make repeated use of comparative adjectives and adverbs. But that does not mean no objective, at least non-arbitrary, judgement is possible. Judgement, in any event, is not impossible, rather inevitable (given the point that no neutral account is on the cards). But that means its grounds can and always must be scrutinised in terms of whatever we hold to be the relevant evidence in the case (and that in turn is always revisable).

It is all very well to pronounce, wistfully, that what we need is a revised framework for the questions or at least a less prejudicial way of posing the questions. But how is that to be achieved? Let me start by developing some points that already surfaced from time to time in our earlier discussions. Several participants drew attention to the difference that differences in scale as well as differences in the contexts of communication may make to knowledge claims, that is to the status of those claims, their modes of justification and their challengeability. It is evidently foolish for us to juxtapose, for comparative purposes, the utterly serious pronouncements of those who purport to appreciate the deepest truths on the most arcane questions on the one hand, and tales told for entertainment or to keep the children quiet on the other. Those who do the pronouncing in the first category may have a distinct status, as pundits, priests, 'shamans', 'philosophers' or 'scientists', many of whom will have socially recognised qualifications or justifications for their position. Some may have fully worked out, explicit, arguments to which they can appeal in support of their claims, if and when these come to be a matter of curious inquiry or of outright challenge. Statements in the second category may be nothing other than playful in intent, but at the same time, we must remain sensitive to the possibility that they are serving profoundly serious purposes.

The pitfalls, snags and obstacles that lie in wait for the unwary should not be underestimated, but nor should they lead to a premature conclusion that the whole project of comparison is strictly impossible. Let me highlight just two examples, from the chapters that follow, that vindicate the comparative ambition and illustrate some of its positive results.

The studies by Humphrey and by Salmond both raise issues concerning spatial cognition, maps, mapping, and navigation. Both authors recognise that there are points of similarity between the ideas and practices of Mongolians or Polynesians on the one hand, and what we ourselves may take for granted in this domain. But both underline striking differences as well. But when what we provisionally label 'maps' turn out, on reflection, to serve very different functions from those we normally associate with such, should our conclusion be, not that we need to revise our ideas about 'mapping', but rather that we are not dealing with anything that can be called a 'map' at all?

The first point to insist on here is that this is not an issue that can be settled a priori, as if there were just a single answer, valid across the board, about the consequences of our encounter with otherness. But the second observation we should make is that that exploration of otherness remains a source of new understandings whichever of the two options we embrace, as between the expansion of our existing concepts or the recognition that they are no longer applicable. After all in the second eventuality, we still have lessons to learn from the encounter. Salmond's discussion of Polynesian navigation shows that it not only meets the criteria that we share – the 'stern test of landfall' in the vast expanses of the Pacific – but also provides an essential component in an alternative framework for thinking about space, time, the environment, the ancestors, persons, 'property' – that is resources – and prosperity. Humphrey in turn shows that while spatial coordinates are present in the work of the Mongolian map-makers she discusses they are not the sole nor even the primary objects of attention, which are better described as exercises in the cosmological placing of the various agents involved, the Emperor to whom the maps are directed, the Banner lords who commissioned them, and those who actually made them, not forgetting the herders who provided the information they contain. To read these objects involves a veritable exercise in translation between the competing languages, that is the conflicting interests and assumptions, of those who created and used them, not that there was ever a lexicon nor even a Rosetta stone that would serve as a guide in the matter – a situation that has more or less strong analogues with those we may say all our contributors encounter.

A second telling set of examples relates to 'mathematics', a field of study that already attracted a lot of attention in our first workshop and that illustrates very strikingly the strain our existing disciplinary categories may be under when confronting heterogeneity and pluralism. Where many accounts of the confrontation between different belief-systems have focussed on religion, or kinship, or alimentary practices, or the body, it is salutary to pursue this agenda in a discipline in which objectivity has often conventionally been assumed to hold unchallenged sway. Yet a moment's reflection is enough to reveal how 'mathematics' too provides grist to our mill. It is easy to see where some attempts to write the history of mathematics have gone wrong, those that took the Euclidean variety as the starting-point and model and dismissed traditions that did not conform to that as worthless. That put paid to the Babylonians, the Indians, the Arabs, the Chinese, the Mayans and everything that ethnography offered by way of indigenous mathematics. But of course most twentieth- and twenty-first-century mathematics does not fit the Euclidean stereotype either.

The historical and ethnographic records present us not just with different, but quite explicit, categories of what counts as 'mathematics', but also with practices that are evidence for varying implicit understandings of numbers and shapes and their manipulation. We put 'mathematics' in scare quotes, to indicate that this involves revising some of our own assumptions but also that the practices in question do indeed imply mathematical skills. To be sure numbers and shapes are often studied not for their own sake but for their symbolic associations, and the calculations undertaken naturally often take into account not just quantities but other interpersonal factors such as past obligations. Modes of counting presuppose notions of the countable and both the modes and the notions exhibit remarkable diversity across the forest and the past. Thus against the grain of some Western authorities, to investigate 'mathematics' in any given human group can mean going into the values it holds dear, its ontological presuppositions, including, as Strathern shows, its ideas about reproduction, propagation and generation. To exclude those other considerations would be to deny the mathematical characteristics of the underlying cognitive capacities. Unity, divisibility, combinability are manipulable quantities and procedures, but ones often fraught with positive or negative connotations with social even cosmological implications. The heterogeneous manifestations of what we take to be mathematical concepts and capacities are brilliantly illustrated here by Vilaça's study of Wari' numeration, by Strathern's of Melanesia, and by McCarty's suggestive comparisons between modern computing and ancient divinatory practices.

It is at that point that we need to recognise the limitations of the notion of mathematics as essentially explicit that we generally take for granted. Note, however, that we still tie 'mathematics' to certain cognitive capacities. It is only if we thought that the planets are alive (which we no longer do) that we would imagine that *they do* mathematics in their circlings. But that leaves what we should say about migrating birds, or bees, or spiders as an open question. And indeed precisely what capacities we attribute to many animal species is not an issue we can take to have been anything like definitively settled.

Another topic that could do with more scrutiny than we gave it in SFSP I is error. It is plausible to claim that all societies, all individuals, are capable of recognising that you can get things wrong. What can that tell us about expectations and assumptions about getting things right? We have to be careful to distinguish between different modalities of that. We should perhaps pay more attention than we did before to the distinction that Tambiah (1968, 1973) made central to his analysis of magical beliefs and practices, namely the contrast between situations where the goal is causal efficacy, and those which aim rather at felicity or appropriateness.

Let me offer an illustration not from Tambiah himself but from Mary Douglas (1966), which allows us to contrast not just indigenous peoples and anthropologists but also divergent anthropological interpretations (cf. Lloyd 2020, 22). Douglas recounted how

> once when a band of !Kung Bushmen had performed their rain rituals, a small cloud appeared on the horizon, grew and darkened. Then rain fell. But the anthropologists who asked if the Bushmen reckoned the rite had produced the rain, were laughed out of court. (Douglas 1966, 73)

Douglas was here drawing on Lorna Marshall (1957), but if we follow up the original report we get a more complex picture. Marshall describes one Rain dance she witnessed:

> One night when the Rain dance was being beautifully danced with a fine precision and vigour of clapping, singing and stamping, which to us suggested fervour, we were

watching it so intently that we had not noticed the sky. The first storm of the season had crept up behind us and suddenly burst over us like a bomb. We asked Gao [her chief native informant] whether he believed the dance had brought the rain. He said that the rain was due to come. The dance had not brought it. (Marshall 1957, 238)

So far so good for Douglas' exposure of the naive belief in the causal efficacy of ritual. Yet elsewhere in that very same article Marshall pointed out a number of rituals danced to 'cure the sick' as well as to 'protect the people … from the spirits of the dead'. There are indeed suggestions in that article that the Bushmen do believe they can control the weather, for example by cutting the throats of particular animals to bring on or to stop rain (Marshall 1957, 239).

This neatly shows how the goal of felicity may overlap with that of efficacy even though the two are still distinct. When patients are sick, they want not just to be comforted but to be cured. True, what counts as being sick or ill in any given group of humans takes you deep into the belief system of that group. Sometimes the illness is as we would say psychological rather than a matter of what we call physical lesions or pathogens. In which case comfort may be the appropriate mode of therapy and so of 'cure'. But those who flock to Lourdes or who went to the shrines of Asclepius in Greco-Roman anti-quity included many who suffered from a whole range of diseases and disabil-ities, blindness, broken limbs, fevers, dysentery, where what counted as a cure was not (just) feeling a bit better but e.g. having your sight restored. In one of the Epidaurus inscriptions the problem the god is consulted about is a lost child. Only the child being found again will count as a result that satisfies the original sense of loss.

For our purposes here the important point to note is that the nature of the judgement of error differs. It may be a matter of the lack of the desired physical effect or one of a failure in that the behaviour was inappropriate, lacking in feli-city. But we have to ask who in the community is in a position to judge? The lack of the restoration of sight – or the failure of the child to reappear – is in the public domain. That the ritual has not been carried out correctly, or even that the wrong ritual has been performed, will often be matters on which special persons, priests, shamans or whoever alone can pronounce – on the basis of knowledge not available to just anyone.

It is especially when things have gone wrong that the question of the nature of the original expectation arises and so too the possibility of challenge to those who claim the authority to decide what should be done. The relevance of this to 'science in the forest' and 'science in the past' is, I trust, clear. 'Science' every-where is fallible, but the modes and consequences of its fallibility vary. Once we factor in the distinction between efficacy and felicity we can avoid one common type of mistake, which was the old canard of treating 'magic' as failed science or failed technology. But positively it opens up new avenues of exploration, of the basis of indigenous knowledge claims, the power structures within the group concerned, the possibilities of challenge and doubt or scepticism. It is never the case, in ancient societies, in contemporary indigenous ones, in the most 'advanced' 'industrialised' ones we are familiar with and live in, that everything works perfectly, that we are entirely in control, with correct understanding reflected in effective behaviour. But one of the advantages of reviewing these issues in the forest and the past is to learn about the variety of human endeavours, to locate our assumptions on that score against the background of the great diversity of the manifestations of human cognitive capacities.

Too often in the past, the really fundamental questions to do with human understanding have been discussed too homogeneously – as if we were dealing with a single and the same issue, of an understander and an understandable, across the board. We were moving away from that in our first workshop, but the present volume is testimony to our sense of the need to do more. Avoiding a certain type of Grand Theory of cognitive capacities (ones that would provide some single explanatory framework for the history of human endeavours to make sense of their surroundings) we implicitly and sometimes explicitly insist that different understandings should be entertained in different contexts. Sometimes when we focus on objects such as maps we have to factor in the considerable diversity in the perceptions and valences of the agents involved. Sometimes our target is rather activities (navigating, counting), or again concepts, such as nature, or whole-part relations, or indeed the world itself. In every case the initial assumptions we might make about what is there to be understood and about how to understand it have to be challenged and revised.

Focussing on different explananda our different studies work towards the appropriate frame within which to advance our understanding of the particular data in question, without assuming that it will be the same frame, or even the same type of frame, linguistic, cultural, political or philosophical, in each case. Allowing for far greater heterogeneity in the types and scale of knowledge claims and practices, in their modes of presentation and justification, and in their implications for interpersonal relations and morality, we hope to be able to do greater justice to the variety of data thrown up by ethnography, history and philosophy of science, and even disputes on the status of Artificial Intelligence (AI).

In the last instance, Blackwell's exceptional fieldwork on AI in Ethiopia prompts him to plead for a less parochial understanding of what 'human' intelligence, taken to be paradigmatic of intelligence as a whole, consists in, with special emphasis on the component of imagination that needs to be included – a view echoed in McCarty's explorations of what we can learn about kinds of intelligence by studying the interplay of humans and machines, the convergence (as he puts it) of human mind and computational affordance. To understand those interactions better he appeals to what at first sight may seem unlikely analogies, with conversation, experimentation (laboratory science) and (as already noted) even divination. Yet all enable us to complexify our understanding of success and failure in inquiry, including the unexpected inter-relational elements in that: we need to go beyond the rigid, reductive, model in play when intelligence is conceived in purely computational terms, to do more justice to the uncertainty, the random, even the 'chaos', we associate already with the non-linear modelling power of the computer.

It is of course not our intention, here or elsewhere in our studies, to open the door to a reversion to the old business of making and policing disciplinary boundaries. Rather, on the contrary, it is to try to get clearer where those disciplines can help one another, not to forge an overarching meta-discipline, but rather to scrutinise where current philosophy of language, philosophy of mind, philosophy of science, our understanding of cognition itself prove to be inadequate. And that is no mere academic matter.

So let me try to summarise some of the key points that may be useful in guiding our methodology, drawing now especially on Jardine's discussion of this fundamental question. What are we doing, exactly, when we investigate sciences that are distant whether in place (the 'forest') or in time (the 'past')? What he offers by way of guidelines as to how to go about this proceeds by a judicious use of a pair of maxims between which there is, as he points out, a certain tension. The 'common ground' maxim focuses on aspects of the material to which we have access that corresponds to some of our own ideas about what counts as scientific, while the 'coherence' maxim (which comes in different forms) concentrates rather on how the beliefs and practices in question relate to what the investigators themselves were interested in. While the application of the first maxim runs the risk of anachronism or anatropism (when we apply the categories that belong to us as 'observers') this can be mitigated by paying due attention to how the actors themselves viewed what they were doing. Jardine exemplifies by citing Rochberg (2016) on Mesopotamian science. Extensive records inform us of their predictions of eclipses and the appearances and disappearances of planets. But to understand what is going on, we need to relate that work to the whole background of the concern with signs and omens that the scribes responsible show. Rochberg's chapter here helpfully adds points to do with the very different understandings of part-whole relations that are often in play in Babylonia where an exploration of the linguistic evidence for mereological themes proves a powerful weapon to bring to light how inappropriate many of the presuppositions in play in Greek-style notions of a 'cosmos' or ours of 'world' may prove to be.

Jardine emphasises that his pair of maxims does not aim to provide a general all-purpose methodology: rather each has always to be applied appropriately to specific cases and fields on a piecemeal basis. But if the aim is to suggest how we gain better access to, and interpret, the data from distant sciences, he further suggests where there are lessons to be learned concerning our own current views in the philosophy of science, for example on the viability of assumptions we may make concerning what we call 'laws of nature'. He thus points the way ahead not just for further work developing what I would call a more ecumenical history of science but also for modifications to our understanding of science and so for the enrichment of philosophy of science itself.

We are all faced with radical otherness, whether we are ancient historians or modern ones or anthropologists. That otherness takes different forms, posing different challenges to our understanding and it does seem important to recognise those differences. Alternative customs are one thing, values another, ontologies, science, mathematics yet others.

In some cases (variety of customs) there is no pressure to suggest there is or should be just the one preferred solution (to how to organise social relations, for example). Strathern's studies, both here and elsewhere (e.g. Strathern 1988, 2019, 2020), exhibit the very different kinds of expectation that may underpin ideas on how this is to be achieved, as too do Vilaça's when she explores how notions of unity and plurality are at work in the domain of morality as much as that of 'mathematics'. Where values are concerned, we would do well to recognise their heterogeneity. No one has a monopoly of the right values to live by. We can and should tolerate others' views. But tolerating others' views does not mean agreeing with all of them and in particular there is a limit to a tolerance of the intolerant.

But what about where ontologies and science are concerned? Here especially we should acknowledge the multidimensionality of the phenomena we are dealing with and allow for pluralism in the explananda. It is not just that some ontological regimes put the emphasis on substances, others on processes. Hugh-Jones's explorations of an Amerindian cosmology dominated by the image of the tube, which have echoes in Strathern's discussion of the images of leakiness and containment in Papua New Guinea, exemplify the range of possibilities with which we have to come to terms. What may start out as 'mythology' is a topic of deep philosophical reflection on the origins of the current dispensation and the rules that govern social life. The body is 'good to think with' and not just in Amazonia but in Western traditions and maybe universally. Yet the reflections it prompts are not given by what anyone might take to be its obvious empirically determined characteristics.

We have to accept that there is no neutral vocabulary in which some unique, univocal account of 'reality' can be expressed. In every natural language terminology exhibits what I call greater or less semantic stretch. But pluralism, even incommensurability, should be seen not as a threat, undermining hope of mutual intelligibility, but rather as an opportunity – for us to learn, not just that our starting assumptions are provisional, but how to overhaul and revise them. There is indeed a double pay-off. We understand others better: and we understand ourselves better too. The goal of the exercise, or one of them, is to allow others their voice and to learn from what we hear them say and see them do.

Of course to make any progress in mutual understanding we must show a willingness to achieve that and there are still plenty of people, even in today's globalising world, who exhibit no inclination to do so. Some people, we have to admit, are simply beyond persuasion. But that should not deter us from continuing our pursuit of the means whereby we can hope to understand one another better. This is the issue we face and the studies that follow accept the challenge to range over millennia and across space in a concerted endeavour to throw some light on these important problems.

Disclosure statement

No potential conflict of interest was reported by the author(s).

References

Douglas, M. 1966. *Purity and Danger*. London: Routledge.

Goody, J. 1977. *The Domestication of the Savage Mind*. Cambridge: Cambridge University Press.

Lloyd, G. E. R. 2020. *Intelligence and Intelligibility: Cross-Cultural Studies of Human Cognitive Experience*. Oxford: Oxford University Press.

Lloyd, G. E. R., and A. Vilaça, eds. 2019. "Science in the Forest, Science in the Past." *HAU: Journal of Ethnographic Theory* 9(1), 36–182.

Lloyd, G. E. R., and A. Vilaça, eds. 2020. *Science in the Forest, Science in the Past*. Chicago, IL.

Marshall, L. 1957. "N'ow." *Africa* 27: 232–240.

Rochberg, F. 2016. *Before Nature: Cuneiform Knowledge and the History of Science*. Cambridge: Cambridge University Press.

Strathern, M. 1988. *The Gender of the Gift*. Berkeley: University of California Press.

Strathern, M. 2019. "A Clash of Ontologies? Time, Law and Science in Papua New Guinea." *HAU: Journal of Ethnographic Theory* 9: 58–74.

Strathern, M. 2020. *Relations: An Anthropological Account*. Durham, NC: Duke University Press.

Tambiah, S. J. 1968. "The Magical Power of Words." *Man* 3: 175–208.

Tambiah, S. J. 1973. "Form and Meaning of Magical Acts: A Point of View." In *Modes of Thought*, edited by R. Horton and R. Finnegan, 199–229. London: Faber and Faber.

Philosophical engagements with distant sciences

Nicholas Jardine

ABSTRACT
This essay considers problems and advantages of philosophical involvement with sciences distant from our own. It endorses the view that, rather than testing of philosophical hypotheses against historical data, this is a matter of interpretation and explication. Guidelines for accessing and interpreting others' sciences are proposed: a common ground maxim, relating to our need to find practices, questions and beliefs that we share with them; and a coherence maxim, relating to the need to do justice to the often alien ways in which others' beliefs and practices may form coherent systems. Examples are then given of ways in which examination of distant sciences, guided by the common ground and coherence maxims, can challenge current philosophical assumptions and enrich our philosophical agenda.

1. Introduction

How, if at all, can philosophy of science profitably engage with the practices and contents of sciences distant from our own? Section 1 briefly addresses the vexed issue of the nature of philosophical engagements with history, endorsing Jutta Schickore's claim that philosophical use of historical (and, I add, anthropological) case studies is a matter of interpretative explication rather than testing of philosophical hypotheses against data. Section 2 is devoted to the problems of interpretation of distant sciences, opening with a defence of the blatant anachronism and/or anatropism (its anthropological counterpart) involved in designating as 'science' beliefs and activities of those who themselves have no concept matching our notion of natural science. I then advance two guidelines for interpretation and communication of distant sciences. One, the 'common ground maxim', urges the interpreter to seek out and explicate practices, questions and beliefs that we share with the subjects of interpretation; the other, the 'coherence maxim', advocates exploration of the ways, often far removed from our own, in which the beliefs and practices at issue form coherent systems. Drawing on works of Daryn Lehoux and Francesca Rochberg on ancient cosmologies, applications of these maxims are illustrated. Section 3 first notes how interpretative engagement with distant sciences can expose our philosophical presuppositions to criticism. Then, building on my own work on Kepler's cosmology and modes of persuasion, it offers the more radical suggestion that reflection on distant sciences and their genres can bring new questions into our philosophical agendas.

2. The nature of philosophical engagements with history

First, some general observations on philosophical dealings with the history of the sciences are in order. As a philosophical naturalist I take it for granted that philosophy of science can and, indeed, should be grounded in the study of the practices and contents of past and present sciences. Further, I endorse views effectively advanced by Schickore on the nature and purposes of engagements of philosophy of science with historical case studies (2018).

To start with, profitable engagement of philosophy with historical cases is not a matter of simple confrontation, analogous to the testing of a theory or hypothesis against observations and experimental outcomes. Rather, it is an iterative (a.k.a. dialectical) process of interpretation and explication involving mutual adjustments of philosophical assumptions in the light of historical accounts and of historical accounts in the light of philosophical assumptions. In particular, philosophical insights are to be gained through detailed explication of the practices and reflections of past practitioners of the sciences, rather than through attention to past general methodologies; Schickore's own historical studies of the roles of experimental replication in the sciences provide fine illustrations of this (2007, 2017). As Schickore notes, such studies serve not only to clarify scientific methods, but can also play a heuristic role, raising 'new points for philosophical discussion about what aspects of scientific practice are indeed epistemic, i.e. conducive to knowledge' (2018, 199).

Cassandra Pinnick and George Gale have argued that the historical adequacy of case studies is at odds with their relevance to philosophy (2000); and Schickore concedes that 'we cannot pursue case studies in the service of criticizing and enriching philosophy of science and at the same time insist that our accounts do justice to the historical situation.'(2018, 200) On this score I have reservations. It is true that historical context will often be of little concern when traditional topics of philosophy of science – induction, explanation, causation, etc. – are addressed. However, historical context is crucial when issues of social epistemology – expertise, authority, testimony, consensus formation – are tackled. As for a measure of 'cherry-picking' and consequent historical anachronism and/or anthropological anatropism, this is, indeed, inevitable in profitable philosophical engagement with distant sciences. However, there are, as will be indicated in the following Section, effective ways in which consequent misrepresentations can often be mitigated.

On the general issue of the nature of philosophical engagements with historical events and episodes, Schickore is, I believe, right to see these as matters of iterative explication rather than scientific testing. But this alone does not suffice to establish their hermeneutic as opposed to scientific nature; for, as she demonstrates in admirable detail in her own studies, engagement of scientific theories with evidence is itself often an iterative process.[1] There are, however, further reasons for maintaining the hermeneutic character of philosophical engagements with the sciences of others.

To start with, philosophical engagements with distant sciences inevitably proceed through interpretation of texts and/or artefacts (and/or, in anthropological cases, utterances). Of course, confrontation of current scientific theories with facts, being a communal enterprise, also requires interpretation of writings and/or utterances; and in special cases involving linguistic difference, interdisciplinary communication, dubious testimony, etc., interpretation may face difficulties. But for the most part such interpretation is unproblematic and can be largely separated from the business of assessment of theory in the light of facts and of facts in the light of theory.

A more radical distinction relating to primary goals may be drawn between scientific engagement with data and philosophical engagement with historical and/or anthropological cases. According to the nomothetic/idiographic distinction, the nomothetic natural sciences have as their primary aim the establishment of general laws, whereas the idiographic human sciences are primarily devoted to the interpretation and explication of particulars: persons, their actions and productions, historical episodes, societies and social changes, etc.[2] It should be emphasized, however, that idiographic and nomothetic approaches are by no means mutually exclusive: nomothetic sciences obviously involve much in the way of interpretation and explanation of individual cases; and idiographic humanities, in explicating individual cases often appeal, whether explicitly or tacitly, to explanatory generalizations.[3] That profitable engagements of philosophy with history are usually idiographic rather than nomothetic is evident, given that insights are generally obtained through detailed interpretation and explication of particular historical episodes rather than through testing of philosophical hypotheses against a plurality of historical instances.

So there are, indeed, solid grounds for endorsing Schickore's insistence on the hermeneutic nature of philosophical engagement with distant sciences.

3. Problems of interpretation of distant sciences

At the outset we must face up to the objection that in writing about 'distant sciences' we impose notions of science and scientific inquiry on the thoughts

---

[1] On such iterative engagement see also Kuukkanen (2018); Scholl (2018).
[2] These terms originate with Windelband (1980 [1894]). On this and allied distinctions between natural sciences and humanities: Schnädelbach (1974), ch. 7; Beiser (2011), chs 8–13.
[3] That idiographic/scientific and nomothetic/humanistic approaches should be seen as complementary rather than antagonistic is urged for history by Wallerstein (1996) and for anthropology by Staley (2012).

and actions of others who generally possess no notion matching or closely analogous to our concept of natural science, and who, moreover, have formed little
or nothing in the way of the social groups, forms of education and institutions
in which our natural sciences are grounded. So the very idea of distant sciences,
while exposing common ground that might not otherwise have been recognized, commits anachronism/anatropism with respect to concepts and forms
of life.[4]

In anachronistically/anatropistically projecting science onto others how are
we to avoid misleading our audiences? The key, I suggest, is open admission of
precisely what we are thus projecting. To do so we may resort to relatively
relaxed notions of scientific practice. For example, with Claude Lévi-Strauss
we might appeal to 'the science of the specific (*la science du concret*)', that is
'the organisation of the perceptible (*sensible*) world in perceptive (*sensible*)
terms' (1966 [1962], 16); or with Geoffrey Lloyd to 'systematic understanding
of a range of natural phenomena, whether or not that understanding is the
result of the self-conscious application of a programme of research governed
by an explicit "scientific method"' (2009, 155); or with Daryn Lehoux to 'projects aimed at understanding, questioning, and testing the natural world'
(Lehoux 2012, 10).

Such expanded notions of science are undoubtedly useful for particular
applications: to totemic systems of plants and animals (Lévi-Strauss); to
ancient Chinese and Greek disciplines (Lloyd); to ancient Roman cosmologies
(Lehoux). However, a requirement of universal applicability would surely
render any such notion too vague to be of use. The issue should rather, I
suggest, be dealt with on a case-by-case basis, explicitly classifying others'
local practices, inquiries, technologies and beliefs as components of a science
whenever they match or are closely analogous to practices, technologies and
beliefs that we consider scientific. On occasion, as just noted, some overall spe-
cification of what is being counted as science may be workable; but often a piecemeal approach will have to suffice.

This approach is in accord with the first of the two general guidelines mentioned above, the common ground maxim advocating the quest for practices,
questions and beliefs that we share with the subjects of our interpretation.
The common ground will often be at the fundamental level of routine practices,
grouping of objects and straightforward observations. Consider, for example,
the classifications involved in local everyday dealings with plants and
animals, so-called folk taxonomies.[5] These may be accounted scientific, as
may local agricultural, culinary and herbal curative practices. Often, however,
the matching will not be in practices that we see as effective or beliefs that
we hold true, but rather at the level of shared questions and forms of

---

[4]On anachronism/anatropism with respect to forms of life, see Peter Winch's classic (1958); Jardine (2000).
[5]See, for example, Berlin (1992); Atran (1993); Junior et al. (2016).

inquiry. So we may take Babylonians' quests for explanation of apparent stellar motions as constituting common scientific ground, regardless of their appeal to divine governance and communication.

In establishing common ground with distant sciences, the quest for explanation of what from our standpoint constitute achievements often plays a major role. On occasion the explanation of such achievements will be obvious to us, as when we take others to recognize seasonal changes in length of the day. But often, especially in the most distant sciences, explanations will be far from obvious, and in seeking them out common ground may be greatly extended. Sometimes such common ground is relatively secure, as with explanation of Babylonian success in prediction of eclipses (Steele 2000). Sometimes, however, it is rather more speculative, as with reconstruction through 'reverse engineering' of the astronomical and calendrical knowledge implied by accurate alignment of prehistoric monuments with the solstices.[6] Pursuit of such explanations is, so I have argued elsewhere, not only an effective means of exploring distant sciences, but also a responsibility that we owe to ourselves and our readerships (Jardine 2019).

But caution is surely in order. For in 'cherry-picking' those of others' actions and beliefs that can be seen as scientific we are apt, at best, to obtain a bricolage of isolated insights and, at worst, to commit gross anachronism and/or anatropism through imposition of our own scientific procedures, agendas and categories. Further anachronism/anatropism may well result from our quest for explanations of what we judge to be their scientific achievements. How are these shortcomings to be avoided? A counter to the risks of gross anachronism and/or anatropism posed by the common ground maxim is provided by our second guideline, the coherence maxim that urges the interpreter to explore the ways, often very far removed from our own, in which the beliefs and practices at issue can be assigned to coherent systems.

What is involved in the coherence maxim? This is a complex issue; but for the arguments that follow a bare summary will suffice. In his account of 'coherence charity' as applied to Roman cosmological texts, Lehoux, following Paul Thagard's analysis of types of coherence, invokes the following: '*explanatory* coherence (when a hypothesis explains, by providing a causal account of a phenomenon)'; '*analogical coherence* (when a hypothesis coheres with a body of theory by being analogous to some accepted part of that body)'; '*deductive* coherence (when a hypothesis coheres because it is deduced from accepted theories, or they are deducible from it)'; '*perceptual* coherence (when we interpret visual stimuli as meaning something about place, size, shape, and type of object)'; and '*conceptual* coherence (when concepts fit together meaningfully)' (Lehoux 2012, 240; Thagard 2000). Lehoux's focus is on the observational and theoretical contents of canonical Roman texts. Where the sources and relics

---

[6]The field of archaeoastronomy, once controversial, is now well established; see Hoskin (2001).

allow access to practical and technological activities, a sixth form of coherence, *deliberative* coherence, comes into play. As Thagard presents this, 'the elements in deliberative coherence are actions and goals, and the primary positive constraint is facilitation: if an action facilitates a goal, then there is a positive constraint between them' (Thagard 2000; cf. Thagard and Millgram 1995). Respect for deliberative coherence is evidently crucial in all engagements with distant sciences that explore practical and technological achievements. And, as we shall see, recognition of yet further types of coherence may on occasion be indispensable for the understanding of distant sciences.

What is the status of the common ground and coherence maxims? Should we go along with Donald Davidson and others who seek conditions of possibility of adequate interpretation (1974)? Or are they merely optional guidelines? Or is some intermediate status best ascribed to them? For present purposes I limit myself to some tentative observations.

There is evident tension between the common ground and coherence maxims. As noted above, the common ground maxim is apt to generate translations and interpretations that assimilate alien actions and beliefs to our own. By contrast, the coherence maxim tends to yield translations and interpretations that reveal the distance of others' patterns of action and belief from our own, and that may well be at odds with our initial findings of common ground. In the vocabulary of translation theory, the coherence maxim 'foreignizes', while the common ground maxim 'domesticates'; and in anthropological terms dating from the 1950s and 60s, the former is 'emic', the latter 'etic'.[7] Consider, for example, Anne Salmond's contribution, in which a foreignizing emic approach told us of the direction of Polynesian navigation by gods, whereas in the course of the subsequent discussion we learned from domesticating etic standpoints how their navigational success could be understood in terms more comprehensible to us.

However, the tension between the maxims is not a matter of mutual abrogation, but rather, at least in their more constructive applications, of what Anna Lowenhaupt Tsing calls 'productive friction' (2005). For our grasp of distant sciences is progressively enhanced through cooperative application of the maxims. For example, Rochberg's work shows how, while finding of common ground in such matters as prediction of eclipses provides us with an entrée to Mesopotamian astronomy, our grasp of its aims and methods is vastly increased through exploration of its coherence with forms of divination, of scribal scholarship and of divine governance (Rochberg 2016). Crucial here, as generally in the study of distant sciences, is the recognition that what we interpret as scientific activity was generally not pursued as an isolated end in itself, but rather as a component and servant of other pursuits.

---

[7]On domestication and foreignization in translation, Venuti (1995) and Eco (2003). On emic and etic approaches, Headland, Pike, and Harris (1990); Jardine (2004).

The relative weights to be attached to these maxims, and to the different forms of coherence covered by the coherence maxim, are dependent on specific features of the texts or utterances under interpretation. For example, when interpreting a text that we take to have scientific content, we tend unthinkingly to impose the standards of presentation that we expect of our own scientific articles and treatises: meticulous description of observations, experimental procedures and their outcomes; rigorous argument and explanation; logical consistency; and, overall, impartial objectivity. Accordingly, we are on the lookout for what we recognize as scientific truths and for explanatory, deductive and conceptual coherence. However, pursuit of these goals in our interpretation of texts presenting distant sciences may be seriously misleading, given that they are often in genres far removed from those of current scientific works.[8] Consider, for example, Cicero's 'Dream of Scipio'.[9] If one focuses strictly on what for us constitutes science, it is seen to present an account of the ordering of the celestial spheres (ultimately derived from Plato's *Timaeus*). But this is to miss out on the richness and coherence of the dream of his journey into the heavens that Cicero has Scipio Africanus relate to his grandson. In Scipio's vision, as Lehoux succinctly summarizes it:

> a divine and eternal god rules the cosmos, the divine and eternal sun rules the other planets, and the divine and eternal human soul rules the human body. Governance, reason, mind, humanity, and cosmos all come together under one set of relationships and responsibilities. The physics is at the same time politics, psychology at the same time ethics and duty. (Lehoux 2012, 184)

So it is only when all aspects of the 'Dream' are appreciated that its forms of coherence become apparent: causal/explanatory with the planets as astrological agents, mediating between heaven and earth in their control of human affairs; analogical in the parallels and harmonies that link heavenly and earthly motions and cycles; and, above all, moral and political coherence.

Switching scale, let us now consider cuneiform astronomy, product of the Babylonian and Assyrian scribal elites, servants of king and state. This is a circa 3000-year tradition, but historians of science have tended to focus on its final centuries, from which we have records of systematic observation, recording and mathematically calculated prediction of celestial phenomena, these being aspects of cuneiform scholarship received into Greek astronomy. However, as Rochberg observes, there were notable continuities throughout the long tradition, such as: 'the focus on signs in heaven and on earth, the prediction of the phenomena valued as signs, observation of cyclical celestial phenomena, response to untoward omens, and increasing attention to interpretation of words used in divinatory contexts' (2016, 35). Rochberg shows how the divinatory reading of celestial signs went along with other

---

[8]On the diverse and often alien genres of ancient scientific works, see Asper (2007); Taub (2017).
[9]Cicero, 'Somnium Scipionis', a part of the final book of *De re publica*.

ominous activities of the learned scribes, notably extispicy, predictions based on inspection of the entrails of sacrificed animals. These activities were in crucial respects scribal: the divinations were readings of divine messages, whether written in the stars or in entrails; and the connections drawn between the ominous signs and the events they foretold often rested on similarities between the words used to describe them (2016, 157). Moreover, the scribal tradition was seen as preserving primordial knowledge vouchsafed by the gods in the remotest antiquity (2016, 65). Here, as with Scipio's dream, the system in which the 'scientific' astronomy is embedded is rich in forms of coherence: perceptual and deductive in the mathematization of celestial and calendrical regularities; causal/explanatory in the communication of divine will through heavenly signs; analogical through reading of signs through terminological similarities; and deliberative through the ominous activities of the scribes as servants of king and state.

Rochberg's account nicely illustrates the dependence on subject matter and purpose of our applications of the maxims of interpretation. If our interest is, for example, in the Greek reception of Babylonian astronomy, then a focus on late cuneiform celestial observation and calculation is in order, and exploration of perceptual and deductive coherence will play central roles alongside indirect application of the common ground maxim. If, however, we are out to relate cuneiform dealings with celestial phenomena to the scribal tradition and its roles in the governance of the state, then attention to analogical and deliberative coherence will be in order.

To summarize, the common ground and coherence maxims are far from being fixed conditions of possibility of access and interpretation. Rather they are guidelines whose weights and modes of application depend upon the subject matters and purposes of our interpretations.

4. Profits from engagement with distant sciences

Let us turn now from problems to prospects. What is to be gained by philosophy from engagement with distant sciences?

One profitable and well-established mode of engagement is that of genealogical critique, the unearthing and exposure to doubt of assumptions, often tacit, inherited from the remote ancestors of current notions. The concept of laws of nature, for example, has been subjected to such critique. Ronald Giere has argued for its inappropriateness, given that 'the whole notion of "laws of nature" is very likely an artefact of circumstances obtaining in the seventeenth century'.[10] And, exploring its classical ancestry, Lehoux has spelled out the ways in which the 'law' in 'law of nature', now a dead metaphor, was then, given belief in divine governance and legislation, literal rather than metaphorical

---

[10]Giere (1999); 89. See also Ruby (1986); the essays in Daston and Stolleis (2008); Van Dyck (2018).

(Lehoux 2006). How then, if at all, can the notion of laws of nature be maintained without positing a ruler beyond or within nature (cf. Beebee 2000)? Or, to take another example, consider the notion of truth in science as objective portrayal of the world. Elsewhere I have argued that this 'realist' notion emerged in astronomy in connection with the aim of mapping and modelling the positions and motions of the heavenly bodies in a manner independent of the observer's standpoint. However, when ascribed to other sciences 'objective portrayal of the world' generally ceases to be literal, becoming a misleading metaphor (Jardine 1988, ch. 9).

More radical is engagement with historically and/or culturally distant sciences directed not at the challenging of current philosophical assumptions, but rather at the enrichment of the philosophy of science by bringing new issues and approaches onto the philosophical agenda.[11] Given the open-ended nature of such engagements with distant sciences, let us call them 'exploratory'.

One area in which exploratory engagement with distant sciences has flourished is philosophical reflection on modes of category formation and application. Anthropological studies of so-called folk taxonomies of plants and animals have raised challenging and hotly contested questions for philosophers and cognitive scientists concerning the extent to which the formation and uses of classifications involve human universals (Steele 2000).

Appeal to mechanisms in the sciences is another topic on which engagement with distant sciences raises new questions. Historical studies of ancient and early modern sciences have revealed a remarkable range of attempts to understand nature through comparisons with man-made machines and contrivances.[12] These studies, along with more recent extensions of the language of mechanism into such fields as economics, hominin genetic/cultural co-evolution and computational neuroscience, not only undermine philosophers' attempts to contrive a universally applicable definition of mechanism, but also raise a series of challenging questions concerning the clarification and justification of the diverse types of metaphors, analogies and models involved in projections of artificial contrivance onto nature.[13]

Ascending from banausic issues of classification and mechanical contrivance to the beauty of theoretical constructs, distant sciences provide ample materials for philosophical reflection. Exemplary on this score are Fernand Hallyn's study of the 'poetics' of early modern cosmologies (1993) and, in a very different vein, Reviel Netz's analyses of the aesthetics of Hellenistic mathematics (2009). Intimately linked with aesthetic appreciation is a field that I believe to be of outstanding philosophical significance, namely the history of rhetorical

---

[11]On uses of history to raise new questions, Koopman (2013).
[12]Berryman (2009); Bertoloni Meli (2006) and (2019).
[13]On extensions of the language of mechanism into economics, Mirowski (1988); into computational neuroscience, Miłkowski (2018); into hominin genetic/cultural co-evolution, Sterelny (2012).

presentation in the sciences. For some 2500 years instruction in rhetoric played major roles in Western education (Kennedy 1980). But in the latter decades of the nineteenth century it was largely eliminated from school and university syllabuses. As Antoine Compagnon suggests, this 'eclipse' occurred 'in the name of new ideologies, the scientific one of objectivity and the romantic one of originality' (1999a, 1251, my transl.). Hand-in-hand with this general waning of rhetoric went the subjugation of the sciences to what Roland Barthes has called 'writing degree zero': colourless, dispassionate, impersonal (1967). However, the past 40 or so years have seen a marked revival of concern with rhetoric (cf. Compagnon 1999b; Fumaroli 1999, 1283–1296). There has been a proliferation of historical/literary works on the, often subdued and surreptitious, rhetoric of scientific articles, treatises and textbooks.[14] In philosophy, so-called argumentation theory has yielded close analyses of techniques of persuasion, and social epistemology has investigated the roles of rhetoric in testimony and achievement of consensus in the sciences.[15] Recent philosophical studies of modes of persuasion in the sciences have, however, paid scant attention to the classical tradition of rhetorical theory and practice.[16] This is, I think, regrettable, given the extraordinary richness and subtlety of reflections on persuasive tactics and their impacts to be found in classical and early-modern works.[17]

In traditional rhetoric oratory was divided into three branches: deliberative (aka legislative), exhorting or dissuading; judicial (aka forensic), accusing or defending; and epideictic (aka ceremonial), praising or blaming. Judicial presentation was widely employed in philosophy and the sciences. For example, much of Seneca's *Natural Questions* proceeds judicially in its raising and responding to objections and in its assessments of the moral standing and reliability of witnesses (Lehoux 2012, ch. 4; Williams 2012, ch. 4). And Alain Segonds and I have shown how in his *Defence of Tycho against Ursus* Kepler presents not Tycho himself, but astronomical hypotheses as his client, defending them against Ursus's calumnies: his sceptical view of them as mere contrivances for 'saving the phenomena'; and his prejudiced and inaccurate account of their history (2008, 169–182). Kepler's work is structured by the standard divisions of a judicial defence. Accordingly, there is an appropriately modest introduction (*exordium*). In each chapter brief introductory material (*propositio, partitio*) is followed by the judicial review (*contentio*). In this, as required in a defence, the *confirmatio* establishes Kepler's own position, while the *refutatio* refutes Ursus's arguments, both directly and through ironic *concessio*, that is, demonstration that Ursus's arguments fail even when contentious claims are conceded to him. In each of these sections of the work the presentation

---

[14]For example, Beer (2004); Bazerman (1988); Gross (1999); Fahnestock (1999).
[15]On argumentation theory: Toulmin (1958); Perelman and Olbrechts-Tyteca (1969 [1952]); and a series of works by Douglas Walton, including his (2013). On rhetoric and social epistemology: Myers (1990); Fuller (1993); and, for an overview, Code (2017).
[16]Notable exceptions are Kienpointer (2001); Walton, Reed, and Macagno (2008), ch. 8.
[17]See, for example, Kennedy (1980); Vasoli (1968); Mack (2011); Skouen and Stark (2015).

follows to the letter the instructions for a judicial defence to be found in the rhetoric textbooks in use at Kepler's alma mater, Tübingen.

In all branches – –deliberative, judicial and epideictic––of classical and early modern rhetoric, three basic strategies were distinguished: *logos*, persuasive argument; *ethos*, appeal to the standing and credibility of the presenter; and *pathos*, appeal to the audience's emotions. As suggested above, accounts of judicial *logos* have much to offer philosophical argumentation theory. As for accounts of the *ethos* and *pathos* of judicial rhetoric, they are, I suggest, potentially a rich resource for the social epistemology of expertise, testimony and formation of consensus in the sciences. In particular, there is surely much to be learned from accounts of *enargeia*, engagement of the audience's attention and assent through lively and moving description (Webb 2009; Plett 2012). As a number of authors have shown, use of vivid description to render audiences 'virtual witnesses' is much in evidence in early-modern presentations of observational and experimental findings.[18] It is also in play in early modern authors' accounts of their routes to discovery, a striking instance being Kepler's *Astronomia nova*, where he explicitly relates his account of his discovery of the form of Mars' orbit to lively narratives of voyages of discovery.[19] Exploratory engagement with the ways in which the various techniques of lively description have been deployed raises, I believe, pressing questions concerning presentation and critical assessment of testimony in the sciences. It raises also more general questions about the roles of empathetic experience and re-enactment in communication.

Enargeia of a different kind figures in the long tradition of works that aim to inspire in the reader sublime visions of the cosmos.[20] Notable examples include the cosmic revelations of Lucretius' *On the Nature of Things* and Seneca's *Natural Questions*, the above-mentioned dream of the world as seen from the heavens recounted by Scipio Africanus in Cicero's *De re publica*, and the description in Kepler's *Somnium* of the earth as viewed by an Icelandic boy guided by a demon to the moon.[21] *Enargeia* in fantastic visions of the cosmos raises compelling questions for coherence theory, in particular concerning the ways in which aesthetic and poetic coherence of a theory relates to other forms of coherence: perceptual, emotional, argumentative, conceptual, analogical, etc. (cf. Thagard 2000, ch. 6).

With respect to both critical and exploratory philosophical engagement with distant sciences, it is worth emphasising the need to substantiate conceptual analysis with 'lexicosemantics', that is, detailed study of the usage and

---

[18]Shapin (1984); Serjeantson (2006); Wintroub (2015).

[19]On *enargeia* in Kepler's *Astronomia nova*, Jardine (2014).

[20]On sublime Roman cosmic visions: Conte (1994); Porter (2016), ch. 5, 'The material sublime'; Shearin (2019). On early-modern fantastic visions of the cosmos: Campbell 1999; Aït-Touati 2011.

[21]Cicero, *De re publica*, VI, 15; Kepler, *Somnium, seu opus posthumum de astronomia lunari* (Frankfurt, 1634). On vivid description in Cicero's 'Dream of Scipio', Lehoux (2012), ch. 8; on the poetics of Kepler's *Somnium*, Reiss (1982), ch. 4.

associations, both literal and figurative, of the relevant vocabulary.[22] As the work of Rochberg and Lehoux amply demonstrates, such lexical investigation is most profitable when it avoids 'cherry-picking' of expressions that seem scientific by our standards, attending rather to the full range of coherent usages. Moreover, insofar as our philosophical concerns are with affective rather than purely conceptual issues––with the rhetorical and aesthetic issues mentioned above, with the emotions that prompt and guide scientific inquiry, etc.––there is surely much to be gained from the burgeoning field of 'sentiment analysis', the use of textual analysis to reveal emotional commitments and aversions (e.g. Liu 2015).

5. Conclusion

We have addressed two main issues: how to gain access to distant sciences; and in what ways distant sciences may be relevant to current philosophy of science. The contributions of distant sciences to philosophy are to be seen, so we have argued, not in terms of testing philosophical hypotheses against historical data, but in terms of interpretation and explication of historical events and episodes. Guidelines for such interpretation designed to facilitate access and promote philosophical relevance have been proposed: the common ground maxim advocating the establishment of concerns, practices and beliefs shared with the subjects of interpretation; and the coherence maxim advocating examination of the ways in which their beliefs and practices form coherent systems. Building on these general reflections, two kinds of philosophical involvement with distant sciences have been discussed and illustrated. Critical engagement, which has been effectively applied to such notions as governance of nature by laws, calls in question current concepts and assumptions through exploration of their ancestry. Exploratory engagement raises new questions for our philosophical agendas. In particular, I have suggested, close attention to the rhetorical practices of distant sciences can enrich the agendas of argumentation theory, coherence theory and social epistemology.

Disclosure statement

No potential conflict of interest was reported by the author(s).

---

[22]A lexicosemantic study of great potential value for philosophical engagement with Roman science is Pellicer 1966.

References

Aït-Touati, F. 2011. *Fictions of the Cosmos: Science and Literature in the Seventeenth Century*, Transl. by S. Emanuel. Chicago, IL: University of Chicago Press.

Asper, M. 2007. *Griechische Wissenschaftstexte: Formen, Funktionen, Differenzierungsgeschichten*. Stuttgart: Franz Steiner Verlag.

Atran, S. 1993. *Cognitive Foundations of Natural History: Towards an Anthropology of Science*. Cambridge: Cambridge University Press.

Barthes, R. 1967 [1953]. *Writing Degree Zero*, Transl. by A. Lavers, and C. Smith. London: Cape.

Bazerman, C. 1988. *Shaping Written Knowledge: The Genre and Activity of the Experimental Article in Science*. Madison, WI: University of Wisconsin Press.

Beebee, H. 2000. "The non-Governing Conception of Laws of Nature." *Philosophy and Phenomenological Research* 61: 571–594.

Beer, G. 2004. *Darwin's Plots: Evolutionary Narrative in Darwin, George Eliot, and Nineteenth-Century Fiction*. 2nd edn. Cambridge: Cambridge University Press.

Beiser, F. 2011. *The German Historicist Tradition*. Oxford: Oxford University Press.

Berlin, B. 1992. *Ethnobiological Classification: Principles of Categorization of Plants and Animals*. Princeton, NJ: Princeton University Press.

Berryman, S. 2009. *The Mechanical Hypothesis in Ancient Greek Natural Philosophy*. Cambridge: Cambridge University Press.

Bertoloni Meli, D. 2006. *Thinking with Objects: The Transformation of Mechanics in the Seventeenth Century*. Baltimore, MD: Johns Hopkins University Press.

Bertoloni Meli, D. 2019. *Mechanics: A Visual, Lexical, and Conceptual History*. Pittsburgh, PA: University of Pittsburgh Press.

Campbell, M. B. 1999. *Wonder and Science: Imagining Worlds in Early Modern Europe*. Ithaca, NY: Cornell University Press.

Code, L. 2017. "Rhetoric and Social Epistemology." In *The Oxford Handbook of Rhetorical Studies*, edited by M. J. MacDonald, 721–732. Oxford: Oxford University Press.

Compagnon, A. 1999a. "La rhétorique à la fin du XIXe siècle (1875-1900)". In Fumaroli 1999a: 1215-1260.

Compagnon, A. 1999b. "La réhabilitation de la rhétorique au XXe siècle". In Fumaroli 1999a: 1261-1282.

Conte, G. B. 1994. "Instructions for a Sublime Reader: Form of the Text and Form of the Addressee in Lucretius' *De Rerum Natura*." In *Genres and Readers*, transl. by G. W. Most, 1–34. Baltimore, MD: Johns Hopkins University Press.

Daston, L., and M. Stolleis, eds. 2008. *Natural Law and Laws of Nature in Early Modern Europe: Jurisprudence, Theology, Moral and Natural Philosophy*. Farnham, Surrey: Ashgate.

Davidson, D. 1974. "On the Very Idea of a Conceptual Scheme." *Proceedings and Addresses of the American Philosophical Association* 47: 5–20.

Eco, U. 2003. *Mouse or Rat: Translation as Negotiation*. London: Weidenfeld and Nicolson.

Fahnestock, J. 1999. *Rhetorical Figures in Science*. Oxford: Oxford University Press.

Fuller, S. 1993. *Philosophy, Rhetoric, and the End of Knowledge*. Madison, WI: University of Wisconsin Press.

Fumaroli, M. 1999. *Histoire de la rhétorique dans l'Europe moderne, 1450-1950*. Paris: PUF.

Giere, R. N. 1999. *Science Without Laws*. Chicago, IL: University of Chicago Press.

Gross, A. G. 1999. *The Rhetoric of Science*. Cambridge, MA: Harvard University Press.

Hallyn, Ferdinand. 1993. *The Poetic Structure of the World: Copernicus and Kepler*. New York: Zone Books.

Headland, T. N., K. L. Pike, and M. Harris. Eds. *Emics and Etics: The Insider/Outsider Debate*. Newbury Park, CA: Sage.

Hoskin, M. A. 2001. *Tombs, Temples and Their Orientations: A New Perspective on Mediterranean Prehistory*. Bognor Regis: Ocarina Books.

Jardine, N. 1988. *The Birth of History and Philosophy of Science: Kepler's A Defense of Tycho Against Ursus with Essays on its Provenance and Significance*. 2nd ed. Cambridge: Cambridge University Press.

Jardine, N. 2000. "Uses and Abuses of Anachronism in the History of the Sciences." *History of Science* 38: 251–270.

Jardine, N. 2004. "Etics and Emics (not to Mention Anemics and Emetics) in the History of the Sciences." *History of Science* 42: 261–278.

Jardine, N. 2014. "Kepler = Koestler: on Empathy and Genre in the History of the Sciences." *Journal for the History of Astronomy* 45: 271–288.

Jardine, N. 2019. "Turning to Ontology in Studies of Distant Sciences." *HAU: Journal of Ethnographic Theory* 9/1: 172–178.

Jardine, N., and A. Segonds. 2008. *La Guerre des Astronomes. Vol2/1: Le Contra Ursum de Jean Kepler*. Paris: Les Belles Lettres.

Junior, W. S. F., P. H. S. Gonçalves, R. F. P. de Lucena, and U. P. Albuquerque. 2016. "Alternative Views of Folk Classification." In *Introduction to Ethnobiology*, edited by U. P. Albuquerque and R. Alves, 123–128. Cham: Springer.

Kennedy, G. A. 1980. *Classical Rhetoric and Its Christian and Secular Tradition From Ancient to Modern Times*. London: Croom Helm.

Kienpointer, M. 2001. ""Modern Revivals of Aristotle's and Cicero's Topics: Toulmin, Perelman, Anscombre/Ducrot"." In *Papers on Grammar: VII, Argumentation and Latin*, edited by A. Bertocchi, M. Maraldi, and A. Orlandini, 17–34. Bologna: CLUEB.

Koopman, C. 2013. *Genealogy as Critique: Foucault and the Problem of Modernity*. Bloomington, IN: Indiana University Press.

Kuukkanen, J.-M. 2018. "Editorial. Can History be Said to Test Philosophy?" *Journal of the Philosophy of History* 12: 183–190.

Lehoux, D. 2006. "Laws of Nature and Natural Laws." *Studies in History and Philosophy of Science* 37: 527–549.

Lehoux, D. 2012. *What Did the Romans Know? An Inquiry Into Science and Worldmaking*. Chicago, IL: Chicago University Press.

Lévi-Strauss, C. 1966 [1962]. *The Savage Mind*. London: Weidenfeld and Nicolson.

Liu, B. 2015. *Sentiment Analysis: Mining Sentiments, Opinions, and Emotions*. Cambridge: Cambridge University Press.

Lloyd, G. E. R. 2009. *Disciplines in the Making*. Oxford: Oxford University Press.

Mack, P. 2011. *A History of Renaissance Rhetoric*. Oxford: Oxford University Press.

Miłkowski, M. 2018. "From Computer Metaphor to Computational Modelling: the Evolution of Computationalism." *Minds and Machines* 28: 515–541.

Mirowski, P. 1988. *Against Mechanism: Protecting Economics from Science*. Totowa, NJ: Rowman and Littlefield.

Myers, G. 1990. *Writing Biology: Texts in the Social Construction of Scientific Knowledge*. Madison, WI: University of Wisconsin Press.

Netz, R. 2009. *Ludic Proof: Greeks Mathematics and the Alexandrian Aesthetic*. Cambridge: Cambridge University Press.

Pellicer, A. 1966. *Natura: Étude sémantique et historique du mot latin.* Paris: Presses Universitaires de France.

Perelman, C., and L. Olbrechts-Tyteca. 1969 [1952]. *The New Rhetoric: A Treatise on Argumentation*, Transl. by J. Wilkinson and P. Weaver. Notre Dame, IN: University of Notre Dame Press.

Pinnick, C., and G. Gale. 2000. "Philosophy of Science and History of Science: a Troubling Interaction." *Journal for General Philosophy of Science* 3: 109–125.

Plett, H. F. 2012. *Enargeia in Classical Antiquity and the Early Modern Age.* Leiden: Brill.

Porter, J. I. 2016. *The Sublime in Antiquity.* Cambridge: Cambridge University Press.

Reiss, T. J. 1982. *The Discourse of Modernity.* Ithaca, NY: Cornell University Press.

Rochberg, F. 2016. *Before Nature: Cuneiform Knowledge and the History of Science.* Chicago, IL: University of Chicago Press.

Ruby, J. E. 1986. "The Origins of Scientific Law." *Journal of the History of Ideas* 46: 341–359.

Schickore, J. 2007. *The Microscope and the Eye: A History of Reflections 1740-1870.* Chicago, IL: University of Chicago Press.

Schickore, J. 2017. *About Method: Experimenters, Snake Venom, and the History of Writing Scientifically.* Chicago, IL: University of Chicago Press.

Schickore, J. 2018. "Explication Work for Science and Philosophy." *Journal of the Philosophy of History* 12: 191–211.

Schnädelbach, H. 1974. *Geschichtsphilosophie nach Hegel. Die Probleme des Historismus.* Freiburg/München: Verlag Karl Alber.

Scholl, R. 2018. "Scenes From a Marriage: on the Confrontation Model of History and Philosophy of Science." *Journal of the Philosophy of History* 12: 212–238.

Serjeantson, R. W. 2006. "Proof and Persuasion." In *The Cambridge History of Science, Vol. 3: Early Modern Science*, edited by K. Park and L. Daston, 132–175. Cambridge: Cambridge University Press.

Shapin, S. 1984. "Pump and Circumstance: Robert Boyle's Literary Technology." *Social Studies of Science* 14: 481–520.

Shearin, W. H. 2019. "Cosmology, Sublimity, and Knowledge in Lucretius' *De rerum natura* and Seneca's *Naturales quaestiones*." In *Cosmos in the Ancient World*, edited by P. S. Horky, 247–269. Cambridge: Cambridge University Press.

Skouen, T., and R. J. Stark, eds. 2015. *Rhetoric and the Early Royal Society: A Sourcebook.* Leiden: Brill.

Staley, R. 2012. "Conversions, Dreams, Defining Aims? Following Boas and Malinowski, Physics and Anthropology, Through Laboratory and Field." *History of Anthropology Newsletter* 39 (2): 3–10.

Steele, J. M. 2000. "Eclipse Prediction in Mesopotamia." *Archive for History of Exact Sciences* 54/5: 421–454.

Sterelny, K. 2012. *The Evolved Apprentice: How Evolution Made Humans Unique.* Cambridge, MA: MIT Press.

Taub, L. C. 2017. *Science Writing in Greco-Roman Antiquty.* Cambridge: Cambridge University Press.

Thagard, P. 2000. *Coherence in Thought and Action.* Cambridge, MA: MIT Press.

Thagard, P., and E. Millgram. 1995. "Inference to the Best Plan: A Coherence Theory of Decision." In *Goal-Driven Learning*, edited by A. Ram and D. B. Leake, 438–454. Cambridge, MA: MIT Press.

Toulmin, S. 1958. *The Uses of Argument.* Cambridge: Cambridge University Press.

Tsing, A. L. 2005. *Friction: An Ethnography of Global Connection.* Princeton, NJ: Princeton University Press.

Van Dyck, M. 2018. "Renaissance Idea of Natural Law." In *Encyclopedia of Renaissance Philosophy*, edited by M. Sgarbi. Cham: Springer. https://link.springer.com/referenceworkentry/10.1007/978-3-319-02848-4_71-1.

Vasoli, C. 1968. *La dialettica e la retorica dell'Umanesimo: "invenzione" e "metodo" nella cultura del XV e XVI siecolo*. Milan: Feltrinelli.

Venuti, L. 1995. *The Translator's Invisibility: A History of Translation*. London: Routledge.

Wallerstein, I. 1996. "History in Search of Science." *Review (Fernand Braudel Center)* 19/1: 11–22.

Walton, D. 2013. *Methods of Argumentation*. Cambridge: Cambridge University Press.

Walton, D., C. Reed, and F. Macagno. 2008. *Argumentation Schemes*. Cambridge: Cambridge University Press.

Webb, R. 2009. *Ekphrasis, Imagination and Persuasion in Ancient Rhetorical Theory and Practice*. Farnham: Ashgate.

Williams, G. D. 2012. *The Cosmic Viewpoint: A Study of Seneca's Natural Questions*. Oxford: Oxford University Press.

Winch, P. 1958. *The Idea of a Social Science and its Relation to Philosophy*. London: Routledge and Kegan Paul.

Windelband, Wilhelm. 1980 [1894]. ""Geschichte und Naturwissenschaft". Transl. G. Oakes, 1924." *History and Theory* 19: 169–185.

Wintroub, M. 2015. "The Looking Glass of Facts: Collecting, Rhetoric and Citing the Self in the Experimental Natural Philosophy of Robert Boyle." In *Rhetoric and the Early Royal Society*, edited by T. Skoen and R. J. Stark, 202–236. Leiden: Brill.

Mongolian map-making as practice

Caroline Humphrey

ABSTRACT

In the late eighteenth to early twentieth centuries Qing officials in China periodically instructed their Mongolian subordinates to make maps of the regions they governed. The Qing were aiming thereby to gain information about the Empire, but the Mongols had other concerns. This article focuses on Mongolian map-making practice and argues that it differed from that prevalent at the time in both China and Russia. Concerned less with conveying practical information than with locating the domain in a cosmological understanding of the world, the Mongol maps were holistic. They were also both 'participatory', involving several different actors in their compilation, and 'relational', concerned to demonstrate the stance of the map-making subject in relation to the map's receiver (the office of the Lifanyuan and ultimately, the Qing Emperor).

Mongolian maps made in the context of Qing imperial projects of governance were surprisingly different, in concepts, techniques and visual imagery, from Chinese maps of the same period. The literature devoted to East Asian cartography has been dominated by the Chinese examples. This paper, which focuses on the Mongol case, attempts to steer a fresh path not only in documenting culturally different practices but also in the approach taken to cartography as an object of study.

The major theme discussed in earlier work was the one that preoccupied Needham, the invention and development of precise cartographic techniques, along with consideration of the influence of European scientific methods. The concern was with 'knowledge' as the core of cartographic activity and with the epistemic-apodictic nature of mapping. Later studies turned to the cultural factors that should be taken into account (Yee 1994) and the political and economic knowledge-gathering aspects of the Qing endeavour (Perdue 1998; Millward 1999). Broadly, maps were to be understood as instrumental-political in character. This article does not ignore either of these perspectives but argues for the need to invoke a third approach, one that is founded on the implications of the practices of map-making. It will be suggested here that Mongol map-

making was both 'participatory' (Crampton 2009), involving several diverse
kinds of actors, and 'relational', since maps were designed for particular recipients or readers. In the Mongol practice, the task of mapping – charting the con-
tours and contents of a territorial surface – was of less concern than the creation
of maps as a 'placing' of the subject, i.e. the domain and its people, in a cosmological vision of the world. The graphic techniques, the approaches taken to
orientation, and the language of inscriptions and reports attached to maps
reveal much about Mongolian spatial thinking.

    To clarify further: the China-focused literature has paid relatively little atten-
tion to the implications of the fact that when the Qing government requested
subject territories to create their own maps and supply them to the capital,
this could not be done according to the cartographic practices known in metro-
politan China. In distant regions such as most of the Mongolian lands, it was
not just that neither the surveying equipment nor the trained specialists were
available (Amelung 2007) but that the very geographical ideas held locally
were different. The Mongols' authorities were left to interpret the imperial
instructions as best they could, using the personnel, techniques and language
at their disposal. The results were extraordinarily varied, and they reflect a
range of graphical traditions, cosmologies, and views on what it was important
to depict that were specific to the Mongols. Much of the academic literature on
Mongolian maps forms a separate body from the Chinese (Heissig 1944, 1961;
Chagdarsurung [Shagdarsüren] 1976; Futaki 2005; Inoue 2014). This article,
while recognizing the situation of the Mongols within the Qing Empire,
seeks to avoid the suggestion that they can only be studied and understood
as 'other' to their imperial masters. The attempt, rather, is to understand the
'science' involved in Mongol map-making. This does not refer to the adoption
of sophisticated Chinese or European ideas and techniques, which in any case
were unavailable to Mongols, but to the intellectual and practical activity
involved in attempts to depict systematically their own understanding of distance, orientation, proportion, and scale in relation to the physical and social
world. The argument is made that where map-making was concerned, the
Mongols, while attempting sporadically to adhere to Qing cartographical
instructions, in practice made use of their own intellectual resources in ways
that simultaneously both conveyed their understanding of geography and cosmology in general and also created 'messages' of self-emplacement to the Qing
authorities. In this sense, the maps have to be understood as relational documents, as 'rhetorical maps', to use the term proposed by Harley (1988) and
Perdue (1998, 275), and as 'performative', as suggested by Crampton (2009).[1]

    In considering the science involved in Mongolian map-making, let me start
with some issues raised by Gell's article 'How to read a map: remarks on the

---

[1]Crampton (2009, 840–843) discusses several senses in which maps have been seen as performative. Among
these, I use the idea of the map as a direct intervention in the conduct of affairs, intervening in particular
between the authors of maps and their recipients.

practical logic of navigation' (1985). The article was *inter alia* a critique of Hallpike's account of 'primitive thinkers', according to which such people are limited to 'practical wayfinding', which is informal, subjective, and based on habit and familiarity, as distinct from technologically advanced kinds of wayfinding, which involve the abstract representation of spatial relations (Gell 1985, 273). Gell rejects this contrast altogether and defends instead a version of the notion, put forward by many geographers, anthropologists, and cognitive scientists, of 'mental maps', i.e. stored information about a terrain, along with some inferential schemes for converting such models into practical decisions and actions. In this view, the mental map used by the experienced native is in one way not so very different from the artefactual map used by a stranger. The key point made by Gell is that in either case, we can only find our way around because we locate our bodies in relation to external coordinates that are unaffected as we move about, and it is only in relation to these coordinates that we hold beliefs about the other kind of location, that of places perceived as relative to our own position (1985, 279).[2] Therefore, he rejects the 'behaviouristic' notion he attributes to Hallpike and also to Bourdieu, that 'practical' space is ego-centric and defined only by coordinates that meet at the agent's own body. Both kinds of knowledge, Gell argued, are necessary in order to know where we are.[3]

In general terms, the Mongolian manuscript maps discussed here[4] – i.e. those created by hand before the appearance of printed maps made according to versions of modern global standards – support Gell's argument. However, they do raise the question, more starkly than is apparent in Gell's article, of what it was that the mapmakers took as the external coordinates, i.e. the spatial relations that I shall call for short 'cosmological', in the sense that they do not change truth value according to place from which they are

---

[2]The example given by Gell is the statement, 'King's Cross is north of here', a statement that depends on the location of 'here'; but the truth value of this statement depends logically on the existence of the spatial proposition of the relation between North and South, which is independent of the location of 'here' (1985, 279).

[3]Hallpike (1986, 342–343) issued a rejoinder to Gell in which he rejected the epithet 'behaviouristic' and the description of his belief, along with that of Bourdieu, as the idea 'that primitive spatial concepts and direction finding are "subject centred" and based on practical mastery'. He argued that Gell had neglected his reference to topological relations, such as proximity, order, inclusion, and separation, which can provide a system of objective orientation without the need for Euclidean or projective representations. For this and Gell's rejoinder, see Man N.S. 21.2: 342–346.

[4]The maps available for this study are by no means all that are preserved in various collections. The Imperial Mongols of the Yuan Dynasty had maps; the defeated Mongol general is recorded as having handed over his maps to the victorious Ming, but they seem not to have survived (see Chagdarsurung [Shagdarsüren] 1976, 345–346 and Herb 1994, 682 for discussion of early Mongol maps and even a globe). Perdue (1998, 279–281) discusses a map made by the Zhungar leader Galdan Tsering in the seventeenth century. Apart from such scant sources, the maps made by Mongols known to modern scholarship date almost entirely from the 1790s onwards. It should be noted, however, that a considerable number of maps not available to me are held in Mongolian collections and some of these may date from earlier periods. I have also not had access to the study by Oyunbeleg (2014) published in China of the Tenri Central Library collection. The maps used for this article are: 9 hand-drawn maps from the Kotwicz archive, dated variously 1805–1912 (Inoue 2014); 182 digitized copies of maps held in the State Library of Berlin, dated 1830s–1930s; and 44 maps held in the Tenri Central Library in Tenri, Japan (Futaki 2005). See details of these holdings in Herb (1994, 682). Courtesy of Dr Tom White I also have access to several undated hand-drawn maps from the Alasha government archives.

stated. If we think comparatively, it is apparent that the world has known many different schemata that can be thought of as universal in this cosmological way. Gell mentions, for example, the 'star courses' used by Pacific navigators and British Admiralty charts; these are very different from one another, but both have the 'essential map-property' of providing non-ego-centred spatial coordinates (1985, 284). One problem posed by Mongolian maps is that they had to negotiate between several such systems known to them, choosing one or another, or alternatively prioritizing one over the other on the same map, or even occasionally giving up on depicting external coordinates altogether – leaving them implicit and requiring the intuition of the map-reader. Another curious feature of many of the Mongolian maps is that 'ego' also often features inside the representation of a territorial space. This invisible 'subjective' agent appears either as the implicit locus of a 'point of view' from which elements of the representation appear or as an inscription in words written as 'I' or 'we' on the map. This paper will make the case that the presence of this 'ego' is not a result of deficient skill in abstract thinking as Hallpike would have it, nor is it primarily to do with wayfinding or practical knowledge. Rather, both of these features, the choice of coordinates and the subject-related features, appear as they do because Mongolian maps were made on behalf of particular princely authorities and addressed to a specific recipient, the Manchu emperor. The maps I will be discussing are not about knowing where we are but about stating where we are to an intended reader.

From the mid-eighteenth century until the early twentieth century, the Qing government in Beijing, on behalf of the emperor, ordered the rulers of their subject Mongolian territories to produce maps of their lands and submit them to the Lifanyuan, the Bureau responsible for administering these regions. In principle, these maps were to be regularly updated but in practice, the supply of maps was desultory and only spiked in the years immediately after an imperial edict.[5] The practice of submitting maps continued under the Chinese Republican and Mongolian socialist governments into the 1920s and 1930s. Almost all of the manuscript maps analysed here were of this kind, having the primary aim of reporting on the boundary, geographical components and situation of the administrative units called 'Banners' (M. *qoshigu, khoshuu*) in the period 1830s–1920s. As wayfinding by locals or strangers was not their purpose, they differ cardinally from the type of map discussed by Gell. So I should clarify that the reason I refer to Gell's paper is because the issues he raises about spatial location and direction are crucial in my view to understanding these Mongolian maps. We could say that all maps, from

---

[5]Some such spike years were 1891, after new mapping regulations were issued in 1890, 1907, when the Qing undertook research on economic resources in Mongolia, and 1911, after the announcement of the New Policy whereby the Qing officially lifted the ban on Chinese settlement in Mongolia and planned to open the country to development. These initiatives along with the calls for maps were a response to the threat of Russian advances (Kamimura 2005, 15–16).

school atlases to the London Underground, are relational to some extent, i.e. made by some author(s) to be perused by some intended readership. The point has often been made that Qing authorities' cartographic projects were set in train to produce maps for themselves, for their own bureaucracies and armies, in order to strengthen their rule over the provinces (Perdue 1998, 266). But the maps discussed here, though a response to the imperial order, were made by local Banner rulers and they, being distant from the capital, depicted location and direction from their own points of view (see Perdue 2003). In terms of power relations, these were maps produced not by masters for themselves, but by subjects for a higher authority. What is interesting and under-researched about these maps is how the Mongols' cartographical devices (direction, coordinates, measurements, labelling, use of colour, etc.) usually considered merely indexical techniques of representing a physical array of objects were used in particular ways: to represent the character and value of the particular territory, to emplace the ruler in it, and sometimes to imply a certain independence of status vis-à-vis the recipient of the map (the Qing state).

The political conditions of production of Mongol maps

The Qing rulers, like those of earlier dynasties, made atlases and maps of their Empire, and being themselves of Manchu hunting/herding/farming stock were aware of the need to cultivate good relations with – and restrain – the potentially volatile nomadic Mongols. In military campaigns, Mongol allies could be an important resource. The Manchus therefore assigned Mongol groups to Banners from which a known number of reserve soldiers could be recruited when necessary. Most of these were to be governed by a single hereditary lord-ruler and ordinary herders were forbidden to stray outside the pastures of their Banner. The Banners were grouped into larger regions, known as Leagues, but since the Qing and later the Chinese Republic were anxious that the Mongols did not join together into unruly threatening conglomerations, the Leagues had little power and the Banners lords were in direct contact with the Imperial administration. This explains why the great majority of the extant maps are maps of individual Banners, floating free on the page, as it were, without extension into neighbouring territories.[6]

Specific maps were requested by the Lifanyuan in order to provide proof of the borders of the Banners, a decision that was taken after the limits of each territory had been decided in principle. The Lifanyuan ordered both

---

[6]However, several of the maps in the Kotwicz collection are of whole Aimags with their constituent Banners (Inoue 2014). This can be explained by the fact that these maps were collected in Da Küriye (later Ulaanbaatar) in 1912, after Outer Mongolia had declared independence from China. The new Mongol government had an interest in consolidating their lands, not dividing them. The Aimag maps seem to have been conceived as the parts of a map of the whole of Mongolia.

League and Banner boundaries to be marked by cairns of stones (M. *obuga*, *oboo*), using an earlier Mongol tradition of building *oboos* on the top of mountains.[7] If no cairns existed on the putative edges of the Banners, new ones were to be erected. The notional line between these cairns would constitute the boundary. In principle, the name of each boundary *oboo* was to be inscribed on the main rock (Chagdarsurung [Shagdarsüren] 1976, 365). However, cairns could be moved, or they might disintegrate into rocky heaps little different from the surrounding landscape. Thus, despite instructions to maintain the *oboos*, considerable doubt about the exact borders reigned well into the twentieth century, which was the cause of numerous disputes between neighbouring Banners. This is no doubt why Constant (2010) argues that the primary reason for the making of maps was to aid in resolving boundary conflicts.

However, a broader examination of Qing map-making policy, as well as examination of the maps, reveals that this cannot have been the only reason – certainly not in the longer historical run. The Kangxi emperor, engaged in the 1690s in military campaigns against the Zhunghar in the far west and having earlier negotiated with the Russians to fix a border in the north, was personally involved in establishing a reliable continent-wide geography and aligning this with geometry, meteorology, and astronomy. He was intensely engaged with the Jesuit missionary scholars who had brought knowledge of measurement of latitude and longitude to China. In this period, the emperor's broad political strategy as well as his own sheer intellectual interest was the main inspiration for map production. As Cams describes (2017, 31–35), the emperor used both European and Chinese cartographic practices to determine precisely the border between the Inner Mongolian lands and Qalqa / Khalkha (Outer Mongolia) (see further below). The ability to make maps for military purposes, ones that accurately reflected the great distances the emperor would lead his Qing armies against the Zhungars, was crucial at this time.

Subsequently, the purpose of ordering the creation of maps shifted. After the conquest of the western regions and border treaties with Russia, although the formerly paramount criterion, assurance of the personal loyalty of Mongolian nobility, did not disappear, the significance of knowledge about the territories themselves gradually grew. It became important to know the dimensions and resources in the administrative the units of the empire and to keep them stable and peaceful. Clarification of boundaries remained salient for governmental purposes. After a dispute over land, a general in the far western Mongolian domain was ordered in 1781 to erect *oboos* at

---

[7] *Oboos* at which the Mongols carried out rites of worship to spirits of the land were marked differently on maps from boundary *oboos* and were often situated in the interior of Banners; nevertheless, I have argued elsewhere (Humphrey 2015) that these *oboos*, which were usually set on mountain tops or high passes, did mark the edge of socially recognized pasturing areas.

the boundary with Qalqa/Khalkha and the officials were ordered to prepare maps. From this time onwards, inter-Banner boundaries too were increasingly officialised. Later, with increasing pressure from Russia, reinforced by the 1860 treaty which gave rights for Russian traders to work inside Qing territory, Chinese officials were sent out in 1867 to produce a gazetteer with appended maps of the Inner Mongolian Banners near the northern frontier.[8] However, even in these areas, as well as in the vast steppes beyond, it was the Mongol authorities who were required to produce maps. While many of their maps mark and name boundary *oboos* according to instructions, others, particularly in the west of Mongolia (Futaki 2005, 30–31), do not delineate a border of any kind; some are content with vague sketches of a mountain range or a river, and yet others depict the Banner as a ruled rectangular outline, when we know that no actual Banner had such a regular geometrical shape. It is evident that borders were not the main preoccupation of the Mongols (see Perdue 1998) (Illustrations 1a and 1b).

These Mongol mapmakers had other concerns. It is clear from the great visual variety of the drawings that they were not the standardized work of officials sent out from Beijing to map Mongol lands. Local artists/draughtsmen must have made the images with their own geographical concepts in mind. To begin to understand what these were, we need to think about how the maps were physically/graphically made, and to do this, we should first take a brief look at the visual languages of mapping available.

The Mongolian maps in the context of Chinese and Tibetan map-making

Little is known about the maps the early Mongols may have used (see footnote 3). During their thirteenth to fourteenth century conquests, they presumably came across the classic Islamic geographies and maps, and later in the seventeenth century, they may have encountered Russian maps (Waugh 2015, 70, 75–79). However, neither of these had a lasting influence. The maps that concern us here were created in the context of several other, different but overlapping, map-making cultures that existed in north and inner Asia during the late Qing. In this section, I look at Mongol practices in the context of Chinese and Tibetan traditions.

In China, despite the existence of mathematical paradigms for map-making since the third century (Li 2018), the use of grids to represent orientation and distance under Persian influence from the Mongol Yuan period (Kauz 2013, 161–162), and the introduction of European ideas of latitude,

---

[8]Zhang Mu and He Qutao *Records of Mongolian herding lands* (蒙古游牧记 *Menggu youmuji*) was issued in 1867 with an appendix containing maps of Inner Mongolian banners. This work was translated by Popov and published in Russia in 1892. http://www.vostlit.info/Texts/Dokumenty/China/XIX/1840-1860/Czan_Mu/frametext1.htm

Illustration 1a. Altai-Urianghai 1912–1914; Horchin Left Banner, 1907; Horchin Right Banner, 1907.

longitude, and Mercator projection by the eighteenth century, practical implementation of these methods was sporadic and had to contend both with inconsistencies between these techniques and with other non-mathematical traditions.[9] The Kangxi Emperor personally compared the Jesuits' measures of latitude using Paris-made instruments with the standard distances in *li* (linear measurements taken with ropes and geometrically corrected) for the distance between Beijing and the Mongolian border town Guisui (now Hohhot). He was disturbed when experiments showed that the two methods did not coincide, concluded that the Chinese standards used in astronomical and cartographic practice were faulty, and in 1698 redefined the length of the *li* by imperial decree so that there were exactly 200 *li* in one degree of terrestrial latitude. Subsequently, other key measurements such as the smaller *chi* were redefined to conform (Cams 2017, 36–37). Many land surveys and maps, using the new *li*, were executed by mixed teams of European missionaries and Qing officials between 1708 and 1717. However, the astronomical and trigonometrical information provided by the Jesuits and its relation to cartography was for a long time restricted to a narrow court circle. And even if it had been made more widely available and well understood, there were not enough specialists trained in surveying nor enough instruments to cover major projects

---

[9]Only one Chinese-made map of the empire using the Mercator projection was produced in the first half of the nineteenth century, and it had little influence and was not published until 1874 (Amelung 2007, 690).

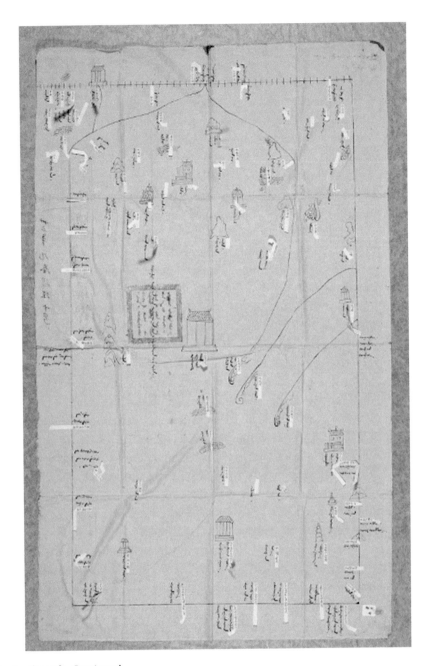

Illustration 1b. Continued.

within China proper, let alone the outer tracts of the empire. Despite recognition of the need for maps, and by the mid-nineteenth century despite exposure to Western seaboard maps, the older traditions and the habit of copying earlier maps generally prevailed. A hybrid exception was the important map of the course of the Yellow River based on Western surveying techniques and delivered to the throne in 1890. It used projection, but the old (incompatible) grid system was employed on the same map. Probably for the first time in China, new 'caterpillar' symbols for depicting mountains appeared, i.e. a view from above rather than the conventional mark depicting mountains from the side; but there was no way of indicating elevation as on European maps (Amelung 2007, 696–698).

Meanwhile, map-making traditions that paid little attention to topographic accuracy continued to flourish. Hermeneutical flexibility was characteristic of Chinese cosmology, which enabled different schema and interpretative practices to coexist, including in cartography (Smith 2013). Incorporated in the multi-stranded Qing culture was a specifically Manchu imperial model. The 'landscape enterprise' of the Qianlong Emperor, as Philippe Forêt (2000, 5–11) calls it, planned the imperial residence of Chengde as a recreation in microcosm of the Buddhist realm of Mount Sumeru. To achieve this scheme, the Emperor used a combination of simple conventional maps, employing a vertical view, pictorial maps that depicted the sacred landscape from the side, looking upwards towards the sacred mountain, and coloured landscape paintings showing the alignment between temples, palaces, and pagodas.

Separate from this, an older Han Chinese non-mathematical tradition continued in the 'descriptive paradigm' of Wei Yuan (1794–1857). In one of his maps, Wei depicted the Qinghai (dark blue sea) as a nebulous spiral hovering an indeterminable distance above the Yellow River, with an explanation written on the page that ' ... even today it does not have the buoyancy to support ships or even a goose feather. Only in bitter winter when the water freezes do Tibetan monks enter into it [walking across the frozen water onto the two islands in the lake] with food for half a year in order to sit and meditate there' (Li 2018, 124). More generally, Wei's concern was with the geopolitical situation of China, the sea-borne threat of the English barbarians, and need for an alternative Chinese geography that represented a dynastic or diachronic greater China extending across Inner Asia, the historical depth being given by use of ancient names and descriptive explanatory texts (Li 2018, 141). Written inscriptions are integral to these maps. The same is true of many of the Mongolian maps – only the kind of text the Mongols added had a different, non-poetic, character, as we will see.

Mongolian mapmakers in the Banners must have known distantly about the Manchu and Chinese enterprises, but their immediate task was to conform to the edicts with the techniques and forms of knowledge they had to hand, which were based on both long-standing indigenous measurements of distance and on Tibetan Buddhist geographical teachings. In contrast to Chinese maps, which were increasingly printed by the nineteenth century, Mongol maps were hand-drawn, using a brush, Indian, or other black ink, and paint or pencils for colouring (Inoue 2014).[10] The Lifanyuan had issued instructions about

---

[10]An exception to hand-drawing is the printed map of Wutaishan, the complex of sacred mountains in northeastern China, made in 1846 by a Mongolian Buddhist lama. This was carved on woodblocks and the prints were then coloured by hand. This map can also be seen as a landscape, in that it takes a single sideways

the symbols to be used – how to draw the sign for a well, an *oboo*, a road, a mountain, and so forth (see list in Chagdarsurung [Shagdarsüren] 1976, 362–364).[11] In 1890 new regulations firmed up earlier standards for mapping area, including the traditional grid, in which each square representing 100 *gazar* (M. 'land', in principle equal to one square *li*) should be drawn as one Chinese *cun* (3.2 cm) on the map (Kamimura and Futaki 2005, 16). The grids were to be aligned with the cardinal directions; they did not refer to latitude/longitude. But these rules could hardly be followed as stipulated. A difficulty was working out how to use these measures and render them on a reasonably sized page when text also had to be fitted in; this meant that the Mongols – even when trying to comply – used different scales, such as 100 *gazar* to 2 *cun*. Many maps left out grids altogether. If the Mongol maps did include a grid – making a bow, as it were, to cartographic basics in the ancient as well as the modern Chinese fashion – they normally provided no information about what the squares were supposed to represent. Furthermore, the Mongol terms used for the three permitted units of distance had their own indigenous meanings, unrelated to the proposed Chinese equivalents: *gazar* was 500 double-steps taken by a man; the *qubi* (M. 'share') was a division of the *gazar*, and the *alda* was the distance between outstretched arms.[12] The accompanying texts in Mongol normally cited distances between boundary *oboos* in *gazar*, *qubi* and *alda*, but some used a variety of other indigenous measurements of distance such as the *num* (M. bow, length of a bowshot[13]) or the *örtege / örtöö* (M. distance between post-relay stations, a fast day's ride, c. 30 km).

The *örtege* was a well-understood 'abstract' unit of measurement that could be used irrespective of the mode of transport. As recently as the 1970s, Mongol herders with experience of ox-caravan travel retained mental maps of the distances and angles of march between a panoply of towns across Inner Asia calculated in *örtege*. They were able to sketch maps on paper of these mental images, marking, for example, the 23 *örtege* between town X and town Y as slashes on the line that represented the route (Humphrey 2020). The ability to do this involved arithmetical and geometrical thinking about geography that surely had a long history among the Mongols, as the post-relay was a highly developed and rather precisely developed technology in the Mongol Empire and the Yuan Dynasty. Daniel Waugh argues (2015, 73–74) that

---

(not vertical) perspective and does not include nomenclature written inside the image. See discussion in Chou (2007).

[11] These instructions were included in a decree entitled in Mongolian *Jarlig-iyar togtagagsan jirug jirugsan kemjiyen-ü dürim basaküü jirug ögülel-ün dürim-ün tuqai*, cited by Chagdarsurung [Shagdarsüren] (1976, 357, but without giving a date).

[12] See Kamimura (2005, 16–19) and Chagdarsurung [Shagdarsüren] (1976, 368) for further discussion of these measures and their putative equivalents in the metric system. According to the latter, the *gazar* was 576 m, the *qubi* was 1/10 of a *gazar*, and the *alda* was 1.6 m.

[13] See Chan (1995) for a discussion of the use of the bowshot as a measure of length in Asia and the Altaic world specifically. In early Mongol texts, the basic unit of measurement was the *alda* (outstretched arms span) and the bowshot feats of heroes were given as amazing distances counted in *alda* (Chan 1995, 36–37).

depicting itineraries was central to many Eurasian cultures of map-making. However, while we could speculate that depicting post-relay, along with military, routes may have been important in early Mongol times, tracing itineraries were not central to the late Qing-era maps, many of which do not indicate roads/routes of any kind. The reason for this absence may be that Mongols did not need roads for travel across the steppes and the relay routes used by officials were of less significance to them than other aspects of the landscape.

With the spread of Tibetan Buddhism in the sixteenth to seventeenth centuries, a different way of thinking about the world had come to pervade Mongolia. This is pertinent to our topic since Buddhist learning formed the educational background not only for the nobles who ordered the maps but also for the scribes who drew them. Akira points out (2005, 13) that the production of maps was a process of negotiation, and he suggests that three actors in the 'power system' were involved: the Emperor of China (or from 1911 the Bogd Khaan of independent Mongolia), the local ruling prince, and the herdsmen on the ground. Of these three, we can see the sovereign (emperor/Bogd Qagan) as the intended ultimate recipient, the local ruler as the authorizing agent for the map, and the herders as the people producing the information about places, distances, names, etc. that went into it. The mapmakers were a fourth set of participants. They were either monks trained in drawing/painting at one of the numerous monasteries that had been founded across Mongol lands, or former lamas, employed on a rotational basis as clerks at the Banner office. Many of their tasks were secular, such as translation of official documents, tax collection, and conduct of the ruler's correspondence, as well as local map-making. But their religious educational background is significant, since it included not only the techniques for painting the ideal landscapes in which deities presided but also the depiction of map-like cosmographical schema of Jambudvipa, the this-world continent of the four arranged symmetrically around Mount Sumeru and often represented by a mandala. Matthew Kapstein argues that although this Indic cosmology was notional and largely imaginary, during the eighteenth century, it became important for certain highly influential lama scholars to determine 'just where on Jambudvipa are we?', the title of his 2011 article. Sumpa Khenpo Yeshe Peljor, a learned monk who belonged culturally between Tibet and Mongolia, was concerned to align the plentifully available real geographical knowledge available to the Tibetans with his description of Jambudvipa. Thus, he included places as far away as Astrakhan, Moscow, and even Sweden in his textual account of the far reaches of the continent. However, what the Tibetans did not do, perhaps out of simple lack of interest, was to employ their skills in mathematics and geometry, used in calculating measurements for cosmic mandala diagrams, in making maps of the practical geography they read about in Sumpa Kenpo's works (Kapstein 2011, 341).

However, the transfer of religious painting techniques to quasi-realist land-scapes did take place among both the Tibetans and the Mongols. Perhaps it would be more correct to say that there was a common visual language that could be used to depict, for example, human scenes illustrating moral precepts in thankas and in map-paintings of monasteries (Harley and Woodward 1994, 607–681). The best-known Mongol example is the work of Marzan Sharav (1868–1939), a former lama who more or less invented the *Monggol zurag* (Mongol painting style), adapting the techniques used in thankas and depictions of pilgrimage sites to secular themes. Sharav was called 'Marzan' (humourous) because his depiction of everyday life (people shearing sheep, defecating, fighting, having a feast, etc.) was unknown, shocking and funny at the time. The *Monggol zurag* style was two-dimensional, adopted variable perspectives as well as a 'bird's eye' view, used colour conventionally rather than realistically, and spread subjects across the canvas to the four corners – all characteristics also found in Mongol Banner maps. Indeed, Sharav's most famous painting, now known as 'One day in Mongolia', has been called a map (Tsultemin 2016) (Illustration 2).

The painting was produced on the order of the Bogd Khaan, the new ruler of Mongolia shortly after the country had gained independence from China. Tsultemin argues (2016) that this work should be understood in relation to similar cartographic manifestations produced elsewhere in Asia when there was

Illustration 2. B. 'Marzan' Sharav, 1912–1915. Mineral paint on cotton, 135 × 170 cm. Zanabazar Museum of Fine Arts, Ulaanbaatar, Mongolia.

anxiety about the unity and survival of a nation. She cites Sankaran Krishna writing about cartography in postcolonial India, suspended in the space between 'former colony' and 'not-yet-nation':

> By cartography I mean more than the technical and scientific mapping of the country. I use the term to refer to representational practices that in various ways have attempted to inscribe something called India and endow that entity with a content, a history, a meaning, and a trajectory. Under such a definition, cartography becomes nothing less than the social and political production of nationality itself. (Tsultemin 2016, 70)

Freed from Qing rule, the Bogd Khaan perhaps had such an aim in mind when he gathered artists and ordered them to go in all four directions and depict everything they saw in their way. It is not known if Sharav was one of these, but his own life had taken him across the country from the Gobi-Altai to the northern landscape of the capital. His painting 'One day' depicts the different geographical zones of the country. At the top is the northern region, known for its rich woods; to the east (right of the picture) are the flat lands of the north-east shown by agricultural fields; while the south-eastern Gobi is indicated by plentiful camels. Map-like in this way, the painting was also highly selective. As Tsultemin points out, the Chinese settlements seen as so threatening by Mongols were left out; the map/painting depicts nothing, but the timeless presence of Khalkha nomads 'hill after hill' (2016, 75).

The distribution of the subjects in 'One Day' is also relevant to the Banner maps I will shortly describe. Seemingly random, the painting was in fact carefully organized. This was done by distributing vignettes to the four corners and relating them *visually* by riders moving diagonally across the scene, and *temporally* by notional sequences (e.g. the bride leaves her home, the bride arrives at the groom's tent), and by making the heart of the painting the Buddhist teaching of the impermanence of life (the depiction of sexual intercourse, birth, and marriage arrayed around a central vignette of lamas conducting a funeral rite) (Tsultemin 2016, 80–82). Read this way, as a map, which Mongol viewers could easily follow from scene to scene, this representation was at the same time far more than a map – in that, at least implicitly, it situated these scenes of life in a cosmology, or perhaps it would be more correct to say, it allowed a cosmological reading of any part of it. Such a 'placing' as I shall now argue was also a feature of those very 'secular' artefacts, the Mongolian Banner maps.

Examples of Mongolian Banner maps

After a long discussion, a close analysis of two Mongol maps brings me back to the questions raised by Gell's article. I start with the issue of the rendering of 'our own position' by the author of the map. The first map (Illustration 3) is of the U̱zümüchin Right Banner made in 1890. Here, rivers and lakes are

Illustration 3. Üzümüchin Right Banner, Shilingol Inner Mongolia, 1890.

seen from above, vertically, while mountains are seen from the side. The overall way to read the map is given by its title written in Mongol vertical script on the reader's right-hand margin. Yet the strange thing is that just over half of the mountains are depicted one way up, and the other half the opposite way. The perspective taken is in effect from *inside* the map. The image makes sense if one imagines an inhabitant journeying diagonally from the map-reader's top right corner down the right bank of the river towards the lake in the centre. This traveller is accompanied by written place-names the right way up for him (her) to read them, and he sees mountains ahead as he would encounter them. However, these names and mountains are upside down to the map-reader. The centre of the map presents respite, a space between the ranks of opposite facing mountains, and here there is a seal stamped over an inscription that provides a key to the whole. 'I,[14] Altanhuyagt, *Qinwang* [ruling prince] of this Banner, have been living and nomadising on both sides of this Balur River for several generations'. The orientation of the mountains suggests that the ruler's pastures spread down to the large lake on the right and along the narrow valley leading to the next river. His worshipped *oboo* is prominently situated behind his main camping area.

    This map has no grid and does not indicate the cardinal directions. However, its rectangular shape is surrounded by 12 regularly placed inscriptions in the margins. These are written from 'outside in', each of them facing inwards to the centre of the map, such that they can only be read by holding the map and turning it round in one's hands. These inscriptions name the neighbouring territories in the 12 cardinal directions, so by reading them, we know that 'south' (*emüne*, 'in front') is at the top of the map from the reader's point of view. Thus, this map bears out Gell's basic argument, that both a local self-emplacement and a relation to abstract coordinates are necessary to understand where one is.

    Yet there is a quirk to this document as a relational object, if one considers that it was made by the Qinwang of the Banner to be presented to the Emperor (via the Lifanyuan). Imagine handing a map to a high person. And imagine receiving it, if, as Smith (2013) argues, your cultural expectations were concerned with loyalty, hierarchy, tributary relations, and ritual correctness. It could be expected that the Mongol princeling would politely present his map in such a way that the high recipient could read it, i.e. the same way up as in the photograph. Now, Üzümüchin is northwest of Beijing. But in the map, the implicit position of the 'I' of the Banner chief residing in his lands is facing the wrong way – not southwards to the capital, but rudely with his back to the Emperor in the south, oriented towards the north at the bottom of the map.

    The question of direction needs to be pursued further. First, it should be noted that the most commonly used Mongol directional words translated as 'south', 'north', etc. also have a possibly older and more primary meaning as 'front/before' (*emün'e*), 'back/behind' (*khoitu*), right hand (*baragun*), and 'left hand' (*zegün*). The Mongol language thus in fact juxtaposes an ego-centric system of spatial coordination with the 'absolute' one the Mongols were equally aware of, referenced by the sun and stars (for discussion of spatial

---

[14]In several of the map inscriptions, 'I' is written as *minii bey-e*, 'my body', which can be related to the bodily stance that lies behind many Mongolian spatial concepts.

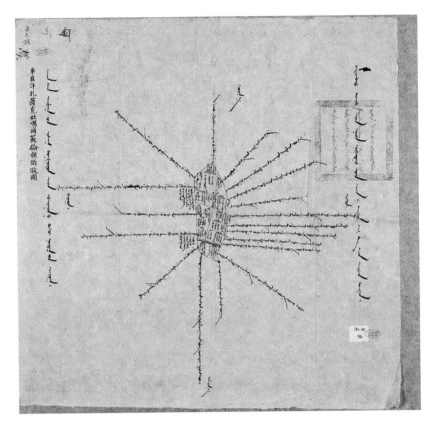

Illustration 4. Lord Dugarsüren's Banner, Setsen Khan Aimag, Outer Mongolia, 1907.

coordination systems and language, see Levinson 2003). That 'front' does not necessarily mean 'south' is seen clearly in Illustration 4.

In this map, the slanting red grid lines dutifully represent the cardinal directions, while the written words 'front', 'back' etc. are clearly at an angle to them. The authors of this map are seeing themselves as facing forwards to the south-east, not the south.[15]

Second, it is perhaps unnecessary to point out that the preoccupations of the mapmakers were not the same as those of officials demanding maps. Yet it was the local people who chose what to put into them and what to omit. Instead of documenting the useful resources, villages, roads, agricultural areas, mines, etc. asked for, some maps were primarily religious, i.e. almost entirely concerned to mark temples and other sacred sites. As can be seen in Illustration 5, apart from rivers painted thickly in red and the ruler's camp, almost everything else of note is a temple or an *oboo*.

---

[15]The Mongols situate themselves and place the doors of their *ger* (yurt) facing forwards/in front. In central Qalqa/Khalkha 'front' (*emüne*) is south or slightly south-east, but for the Western Mongols 'front' veers round to east. A distinction between fixed and moveable objects is made in modern Mongolian. A herder might say, 'My herds are pasturing in front', but not necessarily mean to the south. However, if he is in north Mongolia standing with his back to the town Mandalgov', he would use *emüne* to describe the town's situation.

Illustration 5. Left Ordos Wang Banner, Inner Mongolia, 1911.

It is now possible to understand better another common type of map, which shows that points I have raised separately here (about ego-orientation, cardinal directions, and astrological/cosmological preoccupations) could actually be combined. Illustration 6 is an example of a genre of 'cosmological self-placement' more commonly found in Outer than Inner Mongolia but known in

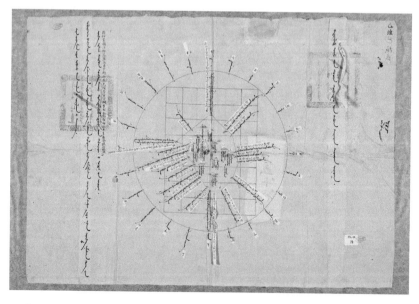

Illustration 6. Banner of Lord Dondubjalbupalmadorji, Tüsiyetü-Khan Aimag, 1907.

both regions. I have chosen to discuss it in detail because Chagdarsurung [Shag-darsüren] (1976) argued, based on his own research in Mongolia, that maps similar to this formed an indigenous Mongol map-making tradition stemming from the Yuan period.

Here, to the reader's left is inscribed the numerous ranks and titles of the lord after whom the Banner is named. To the right is the date. The Banner is an irregular shape in the middle of a red-coloured grid, which is aligned with a set of coordinates arranged in a circle. These are the 12 animals of the cycle of years, with 'horse' at the bottom and 'mouse' opposite at the top. The animal cycle was of course known in China, but it was not used in cartography. In Mongolia, however, as Chagdarsurung describes (1976, 349), the zodiac animal signs were made to correspond not only with time measures (years, months, hours in the day) but also with the four cardinal directions. The latter were subdivided geometrically by the intervening lines, which were called the 8 'colours', the 8 'angles', and the 4 'corners', giving 24 directions in all. These were indicated by their names written centrifugally on maps.[16] In this system, the horse sign represents astronomical south measured by the sun, and thus indexes not a relative but an invariant, 'absolute' system. If this 'zodiac wheel model' is a template that Mongol mapmakers had in mind, it can be seen that the 12 marginal inscriptions in Illustration 3, which are also written centrifugally, follow the same basic pattern. In Illustration 6, Lord Dondubjalbupalmadorji is placing himself and his Banner centrally in a set of abstract coordinates. This seems to be the main concern, since very little is depicted inside the Banner apart from a few hills beneath which are inscribed the names of various pastures (called 'lands' *gazar*). As a relational object, this map is more respectful than the map in Illustration 3. First of all, it appears to conform to the 1890 regulations mentioned earlier: Lord Dondub places his landscape within a grid as instructed, the mountains are all one way up, the 'lands' face south in the 'horse' direction in which the Emperor resides, and strips of paper with translations in Chinese are added for the inscriptions. In a way, this map has been made legible for its high recipient. Yet translating it into anything useful for a government must have presented a puzzling task. No visual indications are given about the surrounding territories. One sees where this Banner is cosmologically, but not in practical geography.

The boundary-marks, the *oboos*, are not much help in this regard. However, the system for designating their geographical relation to *one another* was sophisticated and precise. Each *oboo* is place-named with an outward-directed inscription in all four directions. As studies of Mongol maps describe (Chagdarsurung [Shagdarsüren] 1976; Kamimura 2005; Futaki 2005), each map was supposed to be accompanied by a handwritten 'locality report' (*nutug-un*

---

[16]The centrifugal writing on Mongol maps was probably a matter of convenience for the scribe, since, written from top downwards, a label could more easily be attached to a point by starting there rather than attempting to end there.

chise) providing details of the orientation and distances between boundary marks, which in principle should yield the dimensions of the Banner as a whole. Chagdarsurung transliterates one such report,[17] from which we see that the units of distance used were a mixture of Mongolian and Chinese measures: the *gazar*, *qubi*, *alda*, and the *chi*. The Manchu official receiving it would be faced with a puzzling document, since the report does not mention whether the '*gazar*' taken was the Mongol 500 double-steps or the Chinese *li*. Many of these reports read less like a number of measurements than a list of enchanted connections. For example:

> From the first White Plateau *oboo* in the joyous northwest to the second Black High Peak *oboo* at the source of the small spring in the direction of the interstice between 'Profit'[18] and 'Snake' to the interstice between 'Pig' and 'Eternity'[19], [is a distance of] 18 *gazar*, 5 *qubi*, and 2 *alda*. And from this to the third *oboo* at the pass to the west of the steep mountain ... . (Chagdarsurung [Shagdarsüren] 1976, 351, trs. C Humphrey)

However, this mapping system operated like a compass and was so important to Mongols that after independence, the circular model of the 24 points was approved by all four Qalqa Aimags and incorporated in 1913 into the Legal Code of the new Mongolian state (Kamimura 2005, 17).[20] A key reason was that the zodiac animals system could be used for orientation by tying it to the cardinal directions, and this enabled a translation between Mongol mapping and the systems required in China. Locally, it could be used on maps either to distinguish from or to conflate with the relational front/back system (in the latter case by marking, for example, *morin emüne züg* – 'horse front/south direction' at the bottom of the map). Another reason, however, is that the symbolic animal signs marked time as well as direction and they were used in astrology by Mongols – rulers and herdsmen alike – on a daily basis to make decisions about their activities. In everyday life, it was far less important to know that the 'horse direction' was due south than to know whether this was (or was not) a good direction in which to undertake a journey on such and such a day, or from which to take a wife.

One further point should be made about the map and report quoted by Chagdarsurung and others like it. They essentially recorded notional journey distances, and this is reflected in the language used in some reports, which uses the word 'going' (*yabu-*) (Futaki 2005, 57–59). 'Going' is implicit in the quotation above from Chagdarsurung [Shagdarsüren] (1976, 351), where the mapmaker operates by 'taking a direction' for travel by picturing the diagonal line

---

[17]The report on the Mergen Duke Cheringwanduyibabudorji's Banner in Sechen Khan Aimag in Outer Mongolia, undated. The model used for the map accompanying this report is very similar to Illustration 4, again marking 'front' and 'back' at an angle from the grid supposedly representing 'south' and 'north'.

[18]*Orulta /orulg*-a, 'profit, income bringer', one of the 4 *zobkis* ('corners') equidistant between two of the four cardinal directions.

[19]*Möngke*, 'eternity', one of the 4 *zobkis*.

[20]These regulations also specified the correspondence between the Mongol and the Chinese units of distance, and the scale to be used for the grid (Kamimura 2005, 17).

between two animals on either side of the circle, and then, moving in that line, measures the distance between one *oboo* and the next. It is unlikely, given that the Mongols' means of transport at this period was the horse, camel, or ox, with the measuring technology of a rope or perhaps pacing out the steps, that map-makers could actually have measured the physical distances – perhaps across mountains or forests – with the degree of precision cited in the reports (down to around 1 m). The written reports, therefore, seem to have a rhetorical character, gesturing to exactitude. However, the practice of calculating orientation by opposite pairs of 24 symbols across a circle did enable a fair degree of precision in locating the relation between objects in the landscape. This wheel-like model must have existed as a mental construct of the Mongols so robust that most maps do not need to depict it on the page, even though their accompanying textual reports refer to it (see reports cited in Futaki 2005). After all, it was an everyday presence for all dwellers in the Mongol *ger* (circular yurt). The roof of the *ger* was supported by a wheel of wooden spokes which Mongols aligned with the 12 animals, and they told the time by seeing where the sun's rays fell in relation to the animal-symbol positions.

I would suggest also that a 'wheel' image of the animal symbols must have operated as a key component of 'mental maps' that were shared by Mongols going about their lives as herders. Living in the domain of the Banner, and being forbidden to pasture outside it, herders would be concerned with routes and distances across it, not the lengths around the rim that the map-makers had to trace. Thinking of traversal across the land,[21] they would be interested in the distances between different kinds of pasture, wells, high mountain passes, monasteries, and other features dispersed across the whole of it. They would not need paper maps for wayfinding.[22] Assuming that everyone could observe the position of the sun and held the animal-symbol directions in their heads, routes could be indicated verbally by instructions (in approximately the same laconic form as that used in the reports attached to maps). Yet the maps, which lay out the relations between places visually, indicate a central way that Mongols thought about territorial spaces, that is, to conceive them holistically. This is implied by the use of the word *nutug*, rather than *qoshigu* (Banner) for most of the maps; *nutug* means 'home territory', an assembly – one could even say, a companionship – of dispersed yet inter-related land resources. In this context, it is worth referring to a map that is an exception to the series so far discussed because it seems not to have been produced to central government order (Illustration 7).[23]

---

[21]This is clearer in the 1925 report of a Buryat Banner quoted by Futaki (2005, 59), which gives each distance in the form 'if one travels from X *oboo* in the X sign direction it is X distance to X *oboo*'. In these reports, directions are common cited in terms of the astrological signs in preference to the ambiguous *emüne*, etc.

[22]According to my informants, even people setting out on long pilgrimages from north Asia to Tibet did not use maps but relied on the local knowledge of guides from place to place along the way.

[23]The map, labelled bottom left, depicts the Yellow River (top left) and channels flowing from it, possibly for irrigation. The district had considerable Chinese settlement, denoted by named squares, and the map may have

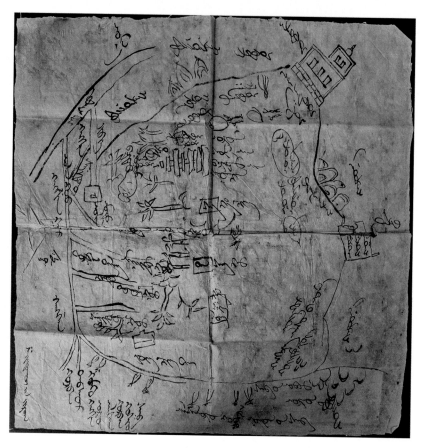

Illustration 7. Shajintohai district of Alasha Banner, Inner Mongolia. Source: Alasha County archives. Date unknown.

What we seem to see in this map is space imagined in a circular manner, as if conceived by someone free to turn 360° and see all around. The cardinal directions (*emüne*, etc.) are clearly marked, but they, off at an angle, have nothing to do with the rectangle of the page. This map, refuses, as it were, the discipline of 'the document'. The writing, the temple, the trees, and the pagoda are depicted from different and even opposite positions. This is dizzying if one imagines a map as having an external 'right way up', but it is understandable if what is being shown is this place from inside, as it is lived in. The scattered disposition of objects and the concern with movements by people and herds diagonally amongst them are features that this map has in common with the Banner maps and with Marzan Sharav's painting. All represent a flattened 'landscape' of widely distributed happenings, with implied 'irregular' (non-rectilinear) journeys from one to another. The Banner maps that are charted in relation

---

been made in connection with land disputes with Mongols. No Mongol government site is visible and the map features rather religious objects such as a monastery, a possibly ruined pagoda (*suburgan*), and sundry *oboos* ranged along the lines that seem to depict a boundary.

to the circle of signs are also saying something else. Each item or event must happen somewhere in geographical space, and the very nature of that point is that it is at the same time an astrological–cosmological location. This is of the greatest concern, since correct alignment of one's astrological self with these spots determines the fate of one's endeavours.

Conclusion

The great majority of Qing-era Mongol maps were produced to order, for a political authority. But despite the hardening of Qing regulations for signage and cartographical standards at the end of the nineteenth century, the huge variety in the maps show that in practice the Mongols continued to follow their own ideas of how to make a map. Very probably, the draughtsmen monks would have known about diverse cross-Asian traditions of cartography, including the Tibetan, the Chinese, and even in northern Banners the Russian. At least some monasteries were host to lamas deeply interested in physical geography of the world beyond the Qing Empire (Kapstein 2011). Possibly some draughtsmen knew about successive Qing Emperors' concern with latitude and longitude, and some are likely to have been aware of traditions of map-making in China that ignored these scientific techniques and chose an altogether different rationale. So, in my view, it would incorrect to see Mongol mapmakers as naïve. A range of possible kinds of the map was available to them: they could choose whether to deploy an ego-oriented representation of landscape or not, which system of coordinates to use, whether to depict a grid, which system of measurements and scale, and whether to make the map schematic or something verging on a realistic landscape painting.

In very general terms, we can say that in making a map of a Banner, the ruler and his draughtsmen were also making a map of a way of life. It is noticeable that the elegantly written titles refer to *qosigu* [*khoshuu*] *nutug* (Banner homeland), the word *nutug* being a multi-scalar notion of 'familiar territory/homeland'. It is noticeable that in the great majority of maps, the most prominent features are mountains, normally individually named and depicted side on, with the worshipped ones carefully marked. As I have described, the ruler often inscribed himself, in his own person and own pastures, into this domain. And commonly, the mountains are depicted facing as they would have looked from the point of view of this ruler moving through his domain (see Illustration 3). As Alfred Gell observed, there is nothing naïve about such self-insertion. He argued that what was required in order to understand where one is (to have a 'mental map' of one's location) is some method by which to relate the bodily position to external coordinates that are independently true (Gell 1985, 279–280). The Mongolian practice of map-making did this in a variety of ways, sometimes more implicitly than explicitly, and usually by having in mind the 'wheel' of symbolic cosmological coordinates that the Buddhist culture taught to be independent and universal. In everyday life, the system of astrological signs was used to coordinate time with spatial orientation; in other words, it determined the auspicious moment linked to the favourable direction for any important activity, journey, or life-event. When embedded in a complex matrix with zodiac-signs, celestial movements and constellations, spirits, elements, etc. this was specialist knowledge encoded in divination manuals (Bauman 2008). But these relational patterns applied to everyone, to groups, great rulers, animals, and omens, and also to non-human occurrences, such as floods or epidemics. For ordinary people, the basic principles of coordination of the system formed a skeletal framework for an extraordinarily holistic way of thinking, which, I have suggested, underlay the making of Mongolian maps. This can be seen most clearly in Marzan Sharav's spatialization of time-specific activities in his 'One Day' map–painting. But in various different ways, the Banner maps show its influence too. Most of them are hybrids that demonstrate the many-layered palimpsest that constitutes 'the local'; but in the practice identified by Chagdarsurung [Shagdarsüren] (1976) with a specifically Mongol tradition, the *nutug* 'homeland' is never shown as a section of land demarcated from the surrounding territory but rather as a unique domain suspended solo between coordinates. This was the case even if the coordinates were taken for granted and not graphically specified. Map-making thus functioned to define a subject of cosmology, perhaps more so than a political entity.

Acknowledgements

I am very grateful to Baasanjav Terbish, Christos Lynteris, and Tom White for their ideas and information, which helped me greatly in writing this article and to the two reviewers for their insightful comments. I would also like to express my gratitude to Dulamjav Amarsaikhan and Uranchimeg Tsultem for their invaluable assistance.

Disclosure statement

No potential conflict of interest was reported by the author(s).

References

Amelung, Iwo. 2007. "New Maps for a Modernizing State: Western Cartographic Knowledge and Its Application in 19th and 20th Century China." In *Graphics and Text in the Production of Technical Knowledge in China*, edited by Francesca Bray, Vera Dorofeeva-Lichtmann, and Georges Métaillé, 685–726. Leiden: Brill.

Bauman, Brian G. 2008. *Divine Knowledge: Buddhist Mathematics According to an Anonymous Manual of Mongolian Astrology and Divination*. Leiden: Brill.

Cams, Mario. 2017. "Blurring the Boundaries: Integrating Techniques of Land Surveying on the Qing's Mongolian Frontier." *East Asian Science, Technology, and Medicine* 46: 25–46.

Chagdarsurung [Shagdarsüren], Ts. 1976. "La Connaissance Géographique et la Carte des Mongols." *Studia Mongolica*, tom III (11), fasc 20, Ulan-Bator: 343–370.

Chan, Hok-lam. 1995. "'The Distance of a Bowshot': Some Remarks on Measurement in the Altaic World." *Journal of Song-Yuan Studies* 25: 29–46.

Chou, Wen-Shing. 2007. "Ineffable Paths: Mapping Wutaishan in Qing Dynasty China." *The Art Bulletin* 89 (1): 108–129. DOI:10.1080/00043079.2007.10786332.

Constant, Frédéric. 2010. "Le gouvernement de la Mongolie sous les Qing: du contrôle sur les hommes à l'administration des territoires." *Bulletin de L'Ecole Française D'Extrême-Orient* 97: 55–89.

Crampton, Jeremy. 2009. "Cartography: Performative, Participatory, Political." *Progress in Human Geography* 33 (6): 840–848.

Forêt, Philippe. 2000. *Mapping Chengde: The Qing Landscape Enterprise*. Honolulu: University of Hawai'i Press.

Futaki, Hiroshi. 2005. "A Description of Boundary Reports (*Nutug-un Cese*) Written in Outer Mongolia in the 1920s." In *Landscapes Reflected in old Mongolian Maps*, edited by Futaki Hiroshi and Akira Kamimura, 27–60. Tokyo: Tokyo University of Foreign Studies.

Gell, Alfred. 1985. "How to Read a Map: Remarks on the Practical Logic of Navigation." *Man N.S* 20 (2): 271–286.

Hallpike, C. R. 1986. "Maps and Wayfinding." *Man N.S* 21 (2): 342–343.

Harley, J. B. 1988. "Maps, Knowledge, and Power." In *The Iconography of Landscape*, edited by D. Cosgrove and S. Daniels, 277–312. Cambridge: Cambridge University Press.

Harley, J. B., and David Woodward. 1994. *The History of Cartography. Volume 2, Book 2, Cartography in the Traditional East and Southeast Asian Societies*. Chicago: University of Chicago Press.

Heissig, Walther. 1944. "Über Mongolische Landkarten." *Monumenta Serica* 9: 123–173.

Heissig, Walther. 1961. *Mongolische Handscriften, Blockdrucke, Landkarten*. Wiesbaden: Franz Steiner.

Herb, G. Henrick. 1994. "Mongolian Cartography." In *The History of Cartography, Volume 2, Book 2, Cartography in the Traditional East and Southeast Asian Societies*, edited by J. B. Harley and David Woodward, 682–688. Chicago: University of Chicago Press.

Humphrey, Caroline. 2015. "'Remote' Areas and Minoritised Spatial Orders at the Russia-Mongolia Border." *Études Mongoles et Sibériennes* 46. DOI:10.4000/emscat.2542.

Humphrey, Caroline. 2020. "'Fast' and 'Slow': Abstract Thinking and 'Real Experience' in Two Mongolian Non-Pastoral Modes of Travel." *Inner Asia* 22: 6–27.

Inoue, Osamu. 2014. "Materials Related to Mongolian Maps and Map Studies Kept at Prof. W. Kotwicz's Private Archive in Cracow." *Rocznik Orientalistyczny LXVII* 1: 116–150.

Kamimura, Akira. 2005. "A Preliminary Analysis of old Mongolian Manuscript Maps: Towards an Understanding of the Mongols' Perception of the Landscape." In *Landscapes Reflected in Old Mongolian Maps*, edited by Futaki Hiroshi and Kamimura Akira, 1–26. Tokyo: Tokyo University of Foreign Studies.

Kapstein, Matthew. 2011. "Just Where on Jambudvipa Are We? New Geographical Knowledge and Old Cosmological Schemes in Eighteenth Century Tibet." In *Forms of Knowledge in Early Modern Asia*, edited by Sheldon Pollock, 336–364. Durham, NC: Duke University Press.

Kauz, Ralph. 2013. "Some Notes on the Geographical and Cartographical Impacts from Persia to China." In *Eurasian Influences on Yuan China*, edited by Morris Rossabi, 159–167. Singapore: Institute of Southeast Asian Studies.

Levinson, Stephen. 2003. *Space in Language and Cognition: Explorations in Cognitive Diversity*. Cambridge: Cambridge University Press.

Li, Man. 2018. "Meaning Behind Maps: Some Remarks on Wei Yuan and His Contribution to Chinese Cartography." *Journal of Asian History* 52 (1): 119–145.

Millward, James. 1999. "Coming into the Map': 'Western Regions', Geography and Cartographic Nomenclature in the Making of Chinese Empire in Xinjiang." *Late Imperial China* 20 (2): 61–98.

Oyunbeleg (ed.). 2014. 蒙古游牧图. Beijing: Beijing Peoples University Publishing House.

Perdue, Peter. 1998. "Boundaries, Maps and Movement: Chinese, Russian and Mongolian Empires in Early Modern Central Eurasia." *The International History Review* 20 (2): 263–286.

Perdue, Peter. 2003. "A Frontier View of Chineseness." In *The Resurgence of East Asia: 500, 150, and 50 Year Perspectives*, edited by Giovanni Arrighi, Takeshi Hamashita, and Mark Selden, 51–77. London and New York: Routledge.

Smith, Richard. 2013. *Mapping China and Managing the World: Culture, Cartography and Cosmology in Late Imperial Times*. New York: Routledge.

Tsultemin, Uranchimeg. 2016. "Cartographic Anxieties in Mongolia: The Bogd Khan's Picture map." *Cross-Current: East Asian History and Culture Review* 21. http://crosscurrents.berkeley.edu/e-journal/issue-21.

Waugh, Daniel C. 2015. "The View from the North: Muscovite Cartography of Inner Asia." *Journal of Asian History* 49 (1–2): 69–95.

Yee, Cordell. 1994. "Chinese Cartography among the Arts: Objectivity, Subjectivity, Representation." In *The History of Cartography*, Volume 2, Book 2, Cartography in the Traditional East and Southeast Asian Societies, edited by J. B. Harley and David Woodward, 128–169. Chicago: University of Chicago Press.

Star canoes, voyaging worlds

Anne Salmond

ABSTRACT

This paper explores the realities of voyaging as understood by early Polynesian navigators. The voyaging world of Kaveia, a contemporary navigator from Outlier Polynesia, as recorded by the anthropologist sailor Mimi George, is juxtaposed with those of Tupaia and Puhoro, eighteenth century navigators from the Society Islands, to examine Polynesian understandings of the sea, canoes and islands, and how long-distance voyaging was accomplished in ancestral times. The renaissance in Polynesian voyaging is also examined, in the context of human harm to the oceanic realm, and modernist understandings of relations between people and the sea are interrogated.

At Borabora in 1818, a half-blind old woman named Ruanui sat with the missionary John Orsmond and chanted the Tahitian creation story. In the beginning, she said, everything emerged from Rua-tupua-nui, the source of growth. When Rua-tupua-nui slept with Atea-ta'o-nui (great expanse), the shooting stars were born, and then the moon, the sun and comets, constellations and the planets.

After the planet Ta'urua (Venus) emerged, he prepared his canoe and sailed west across the sky where he met his wife Rua-o-mere (Capricornus), and they had Maunu'ura (Mars). Maunu'ura sailed south in his canoe and met his wife Apu-o-te-ra'i (vault of the sky), and they had Ta'urua (Formalhault), the steersman of Atutahi's canoe (Atutahi / Pisces Australis). When Atutahi sailed to the West, he came together with Tu-i-te-moana-'urifa (Hydra), and other stars were born.

As each star emerged, they sailed across the skies on their canoes, and other stars were created. The curved arcs of the layered skies were raised up on *pou*, or star pillars,[1] and when the kings of the chiefs of the earth and the kings of the

---

[1] These pou were each named and described in the chant; see Henry (1907).

skies were born, they each had their own star whose names were given to their *marae*, or ceremonial centres. At this time,

> there was prayer in the moving, rolling ocean, the sea was the great marae of the world….Tumu (the rock of foundation) was the husband, Papa (the earth) was the wife, and Oro-pa'a was born to them who still lives in the ocean. He lies with his head upwards, the white foaming breakers are his jaws, he swallows whole fleets of people, he does not spare princes. (Henry 1928, 355)

In this vast oceanic realm, islands were fish, hauled up out of the ocean by the first voyaging ancestors, including Maui, Hiro and Rata. The first dorsal fin of the fish known as Tahiti was its highest mountain, while its second fin was the peninsula Tahiti-iti (Henry 1928, 437–443). Each district had its own mountain, cape and *marae*, with a star ('*avei'a*) standing above it. When the creator ancestor Ta'aroa turned himself into the first canoe, his daughters hauled it to the *marae*, where it became the first *fare atua* (god house): 'The backbone was the ridgepole, the ribs the supporters of the god's house, the breast-bone the capping of the roof, the thigh bones the carved ornaments around the god's house' (Henry 1928, 426).

According to these accounts, land and sea, plants and animals, stars and winds, canoes, god's houses and people were all linked together in one vast kin network. Particular beings could appear in the everyday world in different guises – Ta'aroa as an ancestor, a canoe or a god house; islands as fish; stars as ancestors, pillars or canoes sailing through the sky; whales and sharks as manifestations of ancestors.

Two hundred years later, it is difficult to grasp the kind of reality that underpins these accounts. In outlier Polynesia,[2] however, where ancestral voyaging traditions have survived into contemporary times, they have been passed on to a new generation of navigators. On Taumako, for instance, one of the Duff Islands in far western Polynesia, the *aliki* (high chief) Kaveia began to train a younger generation, including an anthropologist named Mimi George who learned from him over 16 years, sailing with him or directed by him on 25 voyages (George 2012, 140).

In the process, George learned a good deal about sailing with *mana* (ancestral power) on an ocean where the winds, stars, the sea itself, the navigators and their canoes are all ancestors. More recently, she has begun to record these experiences. In Taumako, according to Kaveia, the founding ancestor, Lata (Rata in Tahiti and Aotearoa New Zealand[3]), built the first voyaging canoe. His powers are used by

---

[2]'Polynesia' is a European term coined by French writer Charles de Brosses in 1756. In this article it is used to refer to the islands between Hawai'i, Rapanui / Easter Island and Aotearoa / New Zealand (and some further west) whose indigenous peoples are closely linked linguistically, culturally and historically, and who today often use this term themselves to express their ancestral kinship.

[3]In ancestral times, Aotearoa (or Te Ika ā Māui, or Te Ahi nō Māui) was often used to refer to the North Island of New Zealand, while Te Wai Pounamu referred to the South Island. In contemporary times, Aotearoa is often used to refer to the New Zealand archipelago; or New Zealand, as it was named by Abel Tasman; or Aotearoa New Zealand together.

Taumako navigators to predict and influence the weather, shift winds and waves, and ensure a safe passage between the islands (an ancestral sailing system known as Te Nohoanga te Matangi).[4] The hull of their outrigger canoes ('Te Vaka o Lata') is exceptionally efficient, stabilising the craft. The wings of their crescent-shaped sails ('Te Laa o Lata') are Lata's arms, reaching up to grasp the wind. The belly of the sail is Lata's belly and his mana powers the vessel, while the navigator becomes a living Lata during the voyage. The end of the yard is stepped into a divot carved into the image of a bird lashed to the top of each bow, and 'Lata's teeth' bite into the base of this image from the hull. This type of sail is very powerful, especially with a side or following wind. Under sail, and properly trimmed, the main hull is submerged, except for the carved bird that helped Lata to build the craft, although the prow may plow into the water in steep swells. Lata is at once an ancestor, a navigator, a canoe and a sail, each powered by ancestral mana (George 2018).

As an anthropologist and experienced sailor, George had to unmoor some of her own ontological assumptions to grasp the 'fail-safe' systems used by Taumako navigators, including calling upon Lata in times of danger (George 2012, 147–151). The Taumako wind compass is centred on Lata and animated by his mana, shaping the world.[5] Lata 'dances' with the wind, shifting particular winds around the horizon in their seasonal cycle, and if need be, modifying their position to ensure a safe path between islands, pushing the winds with ancestral power. During a voyage, when Kaveia became Lata, he would sit calmly at the entryway to the cabin on his canoe, working with winds, currents and the stars as the destination island floated towards them. After his death, he appeared on board his canoe to guide those whom he had trained on their voyages, as George vividly describes (2018, 401–404).

According to George, during their voyages together, Kaveia's meteorological forecasts were invariably accurate (2018, 404). She recounts the efficacy of his use of lime-tipped 'weather sticks' to shift particular winds (within their seasonal range) and control waves and rain (2018, 397, 401). Taumako navigators know how to read intricate swell patterns in the ocean (George 2012, 159–165), bouncing off particular islands, the sequences (or 'aveïa') of rising and setting stars, the rising and setting of the sun, the reflection of islands and lagoons on clouds, and migrating and homing birds to find a destination island. George also gives

---

[4]Many thanks to Mimi George for this note.

[5]George (2012, 145–147). See Pyrek (2011) for a critique of George's account of the Taumako 'wind compass,' based on interview notes taken during 5 months of fieldwork on Taumako by her supervisor, Richard Feinberg. Pyrek argues that Taumako navigation is not 'scientific' but 'intuitive' (77–78), claiming that George's earlier accounts of the 'wind compass' were unduly precise as to the direction of the winds in the 'compass,' while exaggerating the role of the winds in Taumako navigation. Neither Feinberg's notes nor Pyrek's account draw on direct experience of sailing with Taumako navigators, nor does Pyrek discuss the role of Lata in Taumako navigation as practiced by Kaveia. Given the depth of George's relationship with Kaveia, this critique seems relatively external and superficial. Pyrek does acknowledge, however, that Taumako navigators generally speak of the winds in terms of the direction from which they blow, a description that evokes ancestral Tahitian accounts that identify particular winds by the holes around the base of the curving, layered skies from which they blew (e.g. in Tyerman 1831, 291–292).

a rich account of *lapa*, flashes like lightening that streak across the ocean from particular islands, each with their own signature; sea paths made visible by Lata's presence that also indicate particular kinds of weather (2012, 150–159). All of these devices Kaveia used to meet 'the stern test of landfall' (Lewis 1972, 43), the empirical proof of navigational skill – finding a destination island rather than perishing in the ocean.

Although George was taught by Kaveia because he wanted the Taumako voyaging tradition recorded for future generations, she found it very difficult to write about those aspects of his navigational practice that exceed the limits of Western rationality. If described in a matter of fact way, these observations invite disbelief, and even ridicule.[6] Indeed, most Western accounts of Polynesian navigation are constrained by assumptions that exclude the reality of ancestral presence and power, and the ability of beings from the ancestral realm to adopt different forms in the realm of everyday world. Arguably, these ontological refusals fundamentally distort our understandings of how the ancestors of Polynesian people were able to explore and settle the Pacific Ocean, given that ancestral beings including the winds, the stars, sea, birds, fish, the navigator him or herself and their canoes, were at the heart of their voyaging system.

Indeed, from the time that the first European ships arrived in Tahiti, one of the islands in the Society Island archipelago, it was clear that Tahitian voyaging was both inspired and driven by ancestral power. At that time Taputapuatea in Ra'iaea, the main marae (ceremonial centre) of 'Oro, the god of fertility and war, was the hub of an extensive voyaging network. Every year, 'Oro's followers, the *arioi*, a society of artists, dancers, musicians, warriors, priests and navigators, carried the gods from island to island, activating the seasonal ritual cycle.

The highest grade of 'arioi, the 'black legs,' men and women distinguished by their red bark-cloth garments and black tattooed legs, enjoyed a privileged lifestyle. Large houses were built to shelter them on their travels, and feasts were staged for their entertainment. They were lavished with gifts, presented with sexual partners and took what they wanted. On their expeditions, fleets of canoes assembled, and travelled under 'Oro's protection to other islands.[7] The missionary William Ellis, who lived in the archipelago during the early nineteenth century, gave a vivid description of the arrival of an 'arioi flotilla:

> [The canoes] advanced towards the land, with their streamers floating in the wind, their drums and flutes sounding, and the Areois, attended by their chief, who acted as their prompter, appeared on a stage erected for the purpose, with their wild distortions of person, antic gestures, painted bodies, and vociferated songs, mingling with the sound of the drum and the flute, the dashing of the sea, and the rolling and breaking of the surf…the whole…presented a ludicrous imposing spectacle.[8]

---

[6]See also an impressive discussion by Teriierooiterai (2013, 47–48).
[7]For an account of the 'arioi, see Salmond (2009, 26–30).
[8]Ellis 1829; quoted in Oliver (1974, 919).

On this occasion, they came to entertain the locals, but when the ʻarioi travelled en masse to the ceremonies at Taputapuatea, their fleets of large, carved canoes carried images of the gods, and pairs of dead men and fish (including sharks and turtles) lay on their prows as sacrifices for ʻOro. As they landed at the marae, drums and conch trumpets sounded, and the bodies of ʻOroʼs sacrificial victims were hung in trees by ropes strung through their heads, or laid as rollers under the keels of the sacred canoes, as they were dragged up the beaches.[9]

In 1769, when James Cookʼs *Endeavour* arrived in Tahiti to observe the Transit of Venus with his Royal Society party of artists and scientists, they soon met Tupaia, a ʻblack legʼ ʻarioi, tahuʼa (high priest) of the Papara district, and a skilled faʼatere (navigator). Tupaia had trained at Taputapuatea, and in about 1760 when Raʼiatea was invaded by warriors from Borabora, a neighbouring island, he fought against them. After the high chief of the island was killed, Tupaia took the high chiefʼs son, a red feather girdle and an image of ʻOro on board ʻOroʼs sacred canoe and carried them to safety in Papara in Tahiti. There he became the lover of Purea, the female ariʼi (high chief) of the district, and tried to help her to install her son as the paramount chief of the island. The other ariʼi joined forces against them, however, and not long before the *Endeavour* arrived at the island, their warriors were defeated in a series of battles.

Although Tupaia had not been exiled, he was now in a marginal position. As an artist and navigator, he was fascinated by the Royal Society party with their watercolours, observatory and astronomical instruments. When Joseph Banks, the wealthy young leader of the Royal Society party, travelled to Moʼorea to observe the Transit of Venus, Tupaia went with him; and on Tahiti, he worked closely with Sydney Parkinson, the shipʼs artist, sketching marae, ʻarioi long-houses, canoes, dancers and musicians, and the chief mourner at a death ritual.[10]

Tupaia also worked with James Cook and Robert Molyneux, the shipʼs master, sharing some of his knowledge about voyaging and navigation, eventually giving them the names of about 130 islands (Salmond 2005, 2008). In Polynesia, where voyaging itineraries were commonly recited as island lists, these itineraries typ-ically included the bearing of the destination island from a particular point on the home island; the succession of stars rising on that bearing throughout the journey (the *rua* or star path); the duration of the journey in *po* or nights; the zenith star (ʻ*aveiʼa*) that marked the position of the destination island; and infor-mation about each island, including its name; size; whether it was low or high; whether or not it had a reef; the location of good harbours; the main foodstuffs produced there; whether or not it was inhabited; whether or not the people were friendly; and the name of its *ariʼi* or high chief.[11]

---

[9]For insights into the links between the seasonal movements of the heavenly bodies, landscapes, ritual sites and Hawaiʼian cosmology and language, see Noyes (2018a, 2018b).

[10]For a detailed account of the life and career of Tupaia, see Salmond (2012).

[11]For a remarkable reconstruction of Tahitian understandings of the cosmos and navigational methods, based on an analysis of early European and Tahitian accounts and linguistic evidence, see Teriierooiterai (2013).

Much of this information (the star paths, for instance) would have been too difficult to share with Cook and Molyneux, or too detailed; and their transcriptions of his lists include only the island names and their bearings from Tahiti.[12] According to Johann Forster's later account, Tupaia 'gave directions for making one according to his account, and always pointed to that part of the heavens, where each isle was situated.'[13] During these conversations, he also shared different aspects of his navigational knowledge, so that Cook's and Molyneux's transcriptions of his lists have only 39 islands in common. Tupaia's lists also include place names from the origin stories, for instance Tumu-papa, a name most likely referring to the creator Tumu (Ta'aroa's phallus) and Papa, the Earth; and various names beginning with Hiti-, evidently drawn from the story of the ancestral voyager Rata (Hiti-teare, Hiti-tautaureva, Hiti-tautaumai, Hiti-poto, Hiti-te-tamaruire, etc.), cosmological features that, from a European point of view, were not real.

On the island list he recorded, James Cook marked the names of twenty islands Tupaia said he had visited, including at least ten in the Society Islands,[14] Niue, two islands in Eastern Samoa, Rurutu (Orurutu), and two in the Cook Islands.[15] This suggests that in the mid-eighteenth century, the 'arioi were voyaging extensively east, west and south from Tahiti. As Molyneux remarked,

> Tobia's Office as Priest has not hinder'd him from travelling which he is very fond of the following extract is from a list of His but sometimes he recollects many more [islands] than is here mentioned…Towbia has seen many of these Islands & has a number more on Tradition that are not here mention'd he is very steady in his account.[16]

When the *Endeavour* set sail from Tahiti, Tupaia decided to go with them, eager to persuade the Europeans to drive his enemies from Ra'iatea, and to visit England. During their voyage around the Society Islands, Cook allowed Tupaia to pilot the ship – a remarkable gesture of confidence. When they set sail and the ship entered

---

[12]Island list transcribed from the second draft of Tupaia's Chart (T2), recorded in James Cook's 'Journal' (T2/C), here as copied by his clerk Richard Orton [Mitchell MS], State Library of New South Wales, Sydney, Safe 1/71; List of islands recorded from Tupaia in Robert Molyneux's Master's Log, National Archives Kew, London, Adm 55/39, 61v (M).

[13]'Tupaya, the most intelligent man that was ever met with by any European navigator in these isles… when on board the Endeavour, gave an account of his navigations and mentioned the names of more than eighty isles which he knew, together with their size and situation, the greater part of which he had visited, and…gave directions for making one according to his account, and always pointed to that part of the heavens, where each isle was situated, mentioning at the same time that it was either larger or smaller than Taheitee, and likewise whether it was high or low, whether it was peopled or not, adding now and then some curious accounts relative to some of them' (Forster 1778, 310–311).

[14]Mytea (Mehti'a), Imao (Mo'orea), Tapooamanu (Maiao), Tethuroa (Teti'aroa), Huiheine (Huahine), Ulietea (Ra'iatea), Otaha (Tahaa), Bolabola (Borabora), Tubai (Tupai), Maurua (Maupiti).

[15]Island list transcribed from the second draft of Tupaia's Chart (T2), recorded in James Cook's 'Journal' (T2/C), here as copied by his clerk Richard Orton [Mitchell MS], State Library of New South Wales, Sydney, Safe 1/71.

[16]See List of islands recorded from Tupaia in Robert Molyneux's Master's Log, National Archives Kew, London, Adm 55/39, 61v - 62r.

a zone of light, variable breezes, Tupaia prayed from the stern windows: 'O Tane, ara mai matai, ara mai matai!' [O Tane, bring me a fair wind!], crying out when the breeze died down, 'Ua riri au!' [I am angry!]. Banks was sceptical about his efforts, however, remarking that Tupaia 'never began till he saw a breeze so near the ship that it generaly reachd her before his prayer was finished' (Banks in Beaglehole 1962, I: 314).

When the ship approached Huahine, Tupaia sent a local man to dive down beneath the keel to see how much water it drew, guiding the *Endeavour* safely through a passage through the reef into the harbour. When he escorted Cook and his party ashore, the high priest stripped to the waist, asking Dr Monkhouse (the ship's surgeon) to do the same as a sign of respect to the gods of the island. Three days later, Tupaia set the course for Ra'iatea, guiding the ship through Te Ava Moa (the sacred pass) through the reef to Taputapuatea, the heart of the 'arioi cult and his home marae, where he failed to convince Cook to attack the Borabora invaders. Afterwards they carried on to Tahaa, Borabora and Hamanino Bay, Tupaia's birthplace on Ra'iatea, a snug cove sheltered by an off-shore reef now occupied by Borabora warriors, where Cook and Banks examined numerous boathouses along the coastline, sheltering large canoes with bellied sides and high peaked sterns, which Tupaia had told them sailed on journeys that took 20 days or more (Banks in Beaglehole 1962, I: 154), but only at certain times of the year.

During this passage through the Society Islands, Tupaia closely worked with Joseph Banks on a chart of the Leeward group, dictating place names and district names which Banks inscribed around the coastlines of the islands.[17] As a result of these exchanges, Cook came to a conclusion about the exploration and settlement of the Pacific that anticipated contemporary scientific findings:

> In these Pahee's [pahi],...these people sail in those seas from Island to Island for several hundred Leagues, the Sun serving them for a compass by day and the Moon and stars by night. When this comes to be prov'd we Shall be no longer at a loss to know how the Islands lying in those Seas came to be people'd, for if the inhabitants of Uleitea have been at Islands laying 2 or 300 Leagues to the westward of them it cannot be doubted but that the inhabitants of those western Islands may have been at others as far to westward of them and so we may trace them from Island to Island quite to the East Indies. (Cook in Beaglehole 1955, 154)

When the *Endeavour* set sail from Hamanino Bay, Tupaia urged James Cook to head to the West, where he said there were plenty of islands that he had visited in a journey that took '10 to 12 days in going thither and 30 or more in coming back' (Cook in Beaglehole 1955, 139fn). Despite Tupaia's entreaties, however, Cook set his course south, intent on resuming his search for Terra Australis.

---

[17]Tupaia, [Chart of the Leeward Society Islands], 1769. British Library, London, BL Add MS 15508, f.16; see also Parsons (2015).

During the long days of their passage, Tupaia often sat in the Great Cabin with Cook and Banks, talking with them and answering their questions. He gave Banks a list of the districts in Tahiti with the number of warriors that each could muster, and tried to share more of his navigational knowledge, telling Cook about the summer westerlies used by island navigators; and Banks about a method of predicting the winds from the shifting curve of the Milky Way.[18] It is likely that these discussions were conducted in a mixture of Tahitian and English, and that during these conversations, both Tupaia's grasp of English and Cook's and Banks's knowledge of Tahitian were extended.

Tupaia's most remarkable contribution, however, was a chart of the islands in the seas around Tahiti, which many scholars have tried to decipher.[19] The latest example, a long, detailed article by Lars Eckstein and Anja Schwarz, is based on a meticulous analysis of three surviving copies of the chart, the surviving island lists, the *Endeavour* logs and journals, modern insights into Cook's cartography, and fragmentary ideas about Polynesian voyaging and navigation (Eckstein and Schwartz 2019; Salmond 2019b). The first version of this chart was drafted after the *Endeavour*'s visit to Rurutu, when Tupaia's direct knowledge of the islands south of Tahiti had been exhausted.[20] It seems that first, Cook, Molyneux or perhaps Isaac Smith, Cook's nephew who assisted him in drafting the ship's charts, or Lieutenant Pickersgill, who completed a copy of Tupaia's chart, laid out the Society islands on the basis of their own charts (to which Tupaia had already contributed). Next, Tupaia placed and named a series of other islands on the chart in relational arrays based on voyaging itineraries that he had memorized during his training and experienced as an 'arioi navigator.

A second iteration of the chart, drafted in Queen Charlotte Sound in Aotearoa, added new information after Tupaia had talked with a local rangatira (chief) about Maori voyaging and geographical knowledge (Eckstein and Schwartz 2019, 78–79). Again, however, none of the Europeans directly involved in this process described exactly how they worked with Tupaia, so many of Eckstein and Schwartz's reconstructions remain speculative. It is clear, however, that for many Maori, at least, the *Endeavour* was Tupaia's ship, and that in many ways, this became an 'arioi voyage (Salmond 2003, 188–189).

From the evidence of the chart and the island lists, it is plain that in Tupaia's time, Tahitian navigators often made long distance voyages. This led Cook and his companions to respect his navigational expertise, and as we have seen, they did not doubt that island navigators were capable of voyages of exploration, or that their ancestors had settled a vast sweep of the Pacific Ocean. They were also impressed by Tupaia's knowledge of the stars, and his ability to orient himself at sea. As Johann Reinhold Forster later wrote:

---

[18]For a detailed account of this traverse from Tahiti and these exchanges, see Salmond (2003, 108–112).
[19]See, for example, Turnbull (1998); Di Piazza and Pearthree (2007); Di Piazza (2010).
[20]Tupaia's Map, 1770. British Library, London, British Library Board BL Add MS 21593.C (T3/B).

Tupaia was so well skilled in this, that wherever they came with the ship during the navigation of nearly a year, previous to the arrival of the Endeavour at Batavia, he could always point out the direction in which Taheitee was situated. (Forster 1778, 509)

Nor were Cook and his companions the only Europeans to be struck by the expertise of Tahitian navigators. Andia y Varela, for instance, a Spanish commander who took a young Tahitian named Puhoro on board the *Jupiter* to Lima in 1774, quickly realized that he had a detailed knowledge of the surrounding islands. When Puhoro saw the ship's pilot drafting a large-scale chart of the islands, he asked what it was, and was told that this drawing represented the islands east of Tahiti. In response, the young man explained that he had visited all of these places, identifying each island by the passages through its reef; reported that there were many more islands in that direction, although he had visited only 18 of them; and told them the number of days it took to sail from one island to another, and which islands had pearls in their lagoons. According to Puhoro, almost all of these places were inhabited.

During the voyage to Lima, Puhoro dictated a list of 15 islands to the east of Tahiti (including most of the NW Tuamotus) and 27 islands to the West, including many of the Society Islands, Atiu and Rarotonga in the Cook Islands, islands in the Marquesas and Fenua Teatea and Ponamu (which Tupaia had not included in his island lists, almost certainly the North and South Islands of New Zealand). As he listed each island, Puhoro made comments about its topography and reefs, its main produce, whether or not the island was inhabited, the ferocity or otherwise of its inhabitants, and how many days it took to sail there from other named islands. Again, the precise sequences of stars to be followed from one island to another were not given, however, in part because it was too difficult to calibrate Tahitian names of stars and constellations with those used by Europeans, and because Tahitian navigation used features such as dark shapes in the sky that were not as significant in European astronomy.[21]

Like Tupaia, Puhoro was an excellent source of knowledge about Tahitian navigational methods, telling the Spaniards about the wind compass used by his compatriots, dictating the names of its 16 points, and explaining how they used the stars to sail to their destinations.[22] According to y Varela:

from him and others, I was able to find out the method by which they navigate on the high seas: which is the following. They have no mariner's compass, but divide the horizon into sixteen parts, taking for the cardinal points those at which the sun rises and sets [...] He knows the direction in which his destination bears: he sees, also, whether he has the wind aft, or on one or other beam, or on the quarter [...]

---

[21]Many thanks to Mimi George for this observation.
[22]For a fascinating discussion of the Tahitian wind compass as dictated by Puhoro to y Varela, with corrected transcriptions into Tahitian, see Teriierooiterai (2013, 129–133).

When the night is a clear one, they steer by the stars; and this is the easiest navigation for them because, these being many [in number], not only do they note by them the bearing on which the several islands with which they are in touch lie, but also the harbours in them, so that, they make straight for the entrance by following the rhumb of the particular star that rises or sets over it; and they hit it off with as much precision as the most expert navigator of civilized nations could achieve. (Corney 1919, 284)

In addition, the Spaniards remarked on the uncanny accuracy with which Puhoro predicted the next day's weather each evening, 'a foreknowledge worthy to be envied, for, in spite of all that our navigators and cosmographers have observed and written about the subject, they have not mastered this accomplishment.'[23]

Invariably, however, as with Tupaia's chart and accounts of Tahitian navigation, ontological as well as linguistic constraints limited these exchanges. A world in which ancestors rode the winds and star ancestors sailed their star canoes across the curved skies; and islands were fish, swimming or floating towards the navigator's craft, which stood still in the ocean; or where navigators became ancestors during their journeys, calling on their mana to summon the winds or calm the sea; or ancestral sharks, whales or birds acted as guardians for a voyaging party – then as now, such accounts were regarded as metaphorical, or 'myths' or 'beliefs' rather than 'matter of fact' accounts of relationships in an oceanic world that enabled the ancestors of Polynesians to invent blue water sailing, and explore and settle the Pacific.

As Teresia Teiwa has observed, the Pacific is the largest geographical feature on the planet (2015) – twelve thousand miles wide across, and ten thousand miles from north to south, covering a third of the earth's surface. Over time, European scepticism about Oceanic realities became a denial that island navigators had the skill to deliberately explore and settle this vast maritime realm, with some scholars arguing that the Pacific must have been settled as the result of accidental drift voyaging. Today, however, the debate has gone full circle. The findings of scientific inquiry, based on archaeological investigations, the DNA of ancient bones and artefacts, and computer simulations that assess the probabilities of different kinds of voyages,[24] have converged with James Cook's conclusion that

if the inhabitants of Uleitea have been at Islands laying 2 or 300 Leagues to the westward of them it cannot be doubted but that the inhabitants of those western Islands may have been at others as far to westward of them and so we may trace them from Island to Island quite to the East Indies. (Cook in Beaglehole 1955, 154)

While particular voyaging sequences are still uncertain, and at times there were migrations north and south, and back to the West, it now seems clear that

[23]For y Varela's account of Tahitian navigation, see Corney (1919, 284–287).
[24]For an excellent recent summary of the scientific evidence on Polynesian origins and voyaging, see Crowe (2018).

the ancestors of Polynesians invented blue water sailing and explored the Pacific from west to east, first from the south coast of China and Taiwan to the Solomon Islands. In a second phase of migration that began perhaps 3,000 years ago, they sailed from east Melanesia from island to island, as far as the west coast of South America. The further east the navigators travelled, the more widely scattered the islands, and the more difficult and dangerous the voyages. Their craft were evidently robust and resilient, their navigational methods capable of guiding them on planned long-distance voyages, and their tales of ancestral exploration based on pioneering maritime achievements.

This conclusion has been reinforced by experimental voyaging, using adapted ancestral navigational methods and craft. This empirical testing began in the 1960s when David Lewis, a New Zealand doctor and sailor, worked with traditional navigators in Outlier Polynesia and Micronesia (Lewis 1972); Bob and Nancy Griffith built two Tuamotu-style voyaging canoes;[25] and Ben Finney, an anthropologist at the University of Hawai'i, worked with the Hawai'ian artist Herb Kane to build the *Hokule'a*,[26] the first of the experimental voyaging canoes. Together, Finney, Kane and Tommy Holmes established the Polynesian Voyaging Society, sparking a great deal of research[27] and a renaissance in non-instrumental navigation. A new generation of voyagers was trained in aspects of seafaring by Mau Piailug from Satawal, a Micronesian navigator, including Nainoa Thompson from Hawai'i, Sir Hekenukumai ['Hec'] Busby, Hoturoa Kerr and Jack Thatcher from Aotearoa New Zealand among others. Thompson created his own star navigation system (with astronomer Will Kyselka), and by these means revivalist navigators have carried out many long-range voyages across the Pacific,[28] including a circumnavigation of the world by the *Hokule'a* in 2014–2017, covering more than 40,000 nautical miles.

During my own work on early cross-cultural encounters in the Pacific, especially the early collaborations between European and island navigators,[29] I had the privilege of meeting Ben Finney, Herb Kane and Mau Piailug in Hawai'i, and later Sir Hekenukumai Busby, Hoturoa Kerr and Jack Thatcher in Aotearoa New Zealand, and became fascinated by their work to revive ancestral voyaging, and navigation by the wind, currents, birds and stars. This led to a series of practical exchanges, with Herb Kane in studying James Cook's ill-fated visit to Hawai'i, and during the 1999 *Endeavour* replica's visit to Kealakekua Bay; with the Waka Tapu project team as they planned a 2012 voyage on *Te Aurere* to Rapa Nui (Easter Island); with Jack Thatcher on *Hine Moana* in Tauranga harbour in 2017, filming for an episode in

---

[25] I am very grateful to Mimi George for this information.

[26] Finney (1986). See also an account of Pacific experimental voyaging in Thompson (2019, 262–296).

[27] E.g. Finney (1998, 1999, 2003); Matamua (2017); the on-line work of the Polynesian Voyaging Society and the Society for Maori Astronomy, Research and Traditions; plus many other sources, including those cited in these notes.

[28] For a detailed, absorbing account of these exchanges and experiments, see Low (2013).

[29] Salmond (2005, 2006, 2008, 2012).

the *Artefact* documentary series on ancestral voyaging; and from 2017 as a member of the Voyage Navigation group that is planning a journey around Aotearoa New Zealand's coastline by a flotilla (including two waka hourua from New Zealand (*Hinemoana* and *Haunui*), *Fa'afaiti* from Tahiti, two tall ships and the *Endeavour* replica) in late 2019 to mark the 250th anniversary of the *Endeavour's* 1769–1770 circumnavigation of New Zealand.

These contemporary voyaging projects have provoked a great deal of reflection about different ways of understanding and traversing the world's largest ocean.[30] Pacific artists, film-makers, curators, scholars and tribal leaders are also inquiring into different relationships with the sea, particularly from ancestral times but also in the present.[31] They are acutely aware of existential threats to the ora (health, prosperity, well-being) of the Pacific Ocean and its inhabitants (non-human as well as human) – rising sea levels and temperatures, acidification, pollution, shifting currents, vast gyres of rubbish, dying coral reefs and drowning islands. In their projects, fundamental questions are being asked about the adequacy of modernist ways of investigating and relating with other living systems and life forms, especially those based on radical splits between Culture and Nature, mind and matter, subject and object, people and 'the environment', and the extractive, destructive practices that arise when other life forms are defined simply as 'things' detached from people, 'property' to be owned and 'resources' to be managed for human purposes.

In Aotearoa New Zealand, this has led to a reassertion of ancestral relationships with land and waterways, as well as the ocean.[32] In recent times, claims against the Crown based on the Treaty of Waitangi (signed between Māori leaders and the Crown in 1840) have had powerful ontological dimensions, arguing that since colonial times, impositions of European conceptions of space, time, persons and property on land, sea and Māori people have fundamentally fractured Māori relations with their ancestral territories, and with each other. In the case of the Whanganui River, for instance, the Whanganui people have argued that they, their ancestors and the river are one, and that their well-being and that of the river are mutually implicated. As one elder lamented to the Waitangi Tribunal:

> It was with huge sadness that we observed dead tuna [eels] and trout along the banks of our awa tupua [ancestral river]. The only thing that is in a state of growth is the algae and slime. Our river is stagnant and dying. The great river flows from the gathering of mountains to the sea. I am the river, the river is me. If I am the river and the river is me – then emphatically, I am dying.[33]

---

[30]Salmond and Salmond (2010); Salmond (2015a, 2017a).
[31]Salmond (2015b, 2015–2016, 2017b, 2018a); in addition, an *Artefact* documentary series on Maori TV, with episode about Pacific voyaging and exploration.
[32]Bambridge (2017).
[33]Tiirama Thomas Hawira in Waitangi Tribunal, *Whanganui River Report* 1999. Wellington, The Waitangi Tribunal (Wai 167), 56.

When the Te Awa Tupua Act (2017) was passed by the New Zealand Parliament to settle the Whanganui Treaty claim, the river was declared to be a legal person with its own life and rights, existentially entangled with Whanganui people.

As Marshall Sahlins observed, 'The [Māori] universe is a gigantic kin, a genealogy…a veritable ontology.' (1985, 195) Whakapapa, the word in Māori for this cosmic kin network, is often described in terms that relate to the ocean. In his work on the Māori 'genealogical method,' for instance,[34] the eminent twentieth century Māori scholar and politician, Sir Apirana Ngata, noted that aho and kaha, words for descent lines, also refer to the cords or lines on which fish or shellfish were strung; while tātai (another term for lineage) also refers to the strands of a net (Ngata 2019). A whakapapa expert might recite a genealogical line as though stringing up the fish in a catch, or trace the lines in a whakapapa as though weaving a net. In speaking of the recent dead, one might invoke the net of Taramainuku, the commander of a star waka that travels through the sky each night, his net sweeping up the wairua (spirits) of those who have died (Wi Repa 1923).

Among the new generation of navigators, innovative adaptations of ancestral voyaging methods and cosmological ideas are emerging, and embodied relations with the ocean and with their voyaging ancestors are taken for granted. Paikea, for instance, the great East Coast ancestor, is at once a whale rider and a whale, a tipua (being from the ancestral realm), and an ancestor. Jack Thatcher, one of Mau Piailug's students and a master navigator descended from Paikea, is leading a group of voyagers who are currently drafting a declaration of the rights of Hine-moana, the Pacific, which they intend to carry on the flotilla that will sail around the Aotearoa New Zealand coastline towards the end of this year to commemorate the 250th anniversary of the arrival of the *Endeavour*.

This reassertion of the Pacific as ancestor provides an explicit alternative to the Western modernist vision of the ocean as a 'wild' maritime expanse, gridded by latitude and longitude, abstracted and emptied of life and people. It rejects the law of the sea, which in the European 'Age of Discovery' took continental conventions about the ownership of the foreshore and seabed, spatial boundaries between territorial waters (now exclusive economic zones) and the high seas, and imposed them on the world's largest ocean (Salmond 2018b).

In the Pacific, however, the law of the sea is failing to safeguard the sea and its inhabitants. A kin-based ontology in which people, fish, winds and stars are intertwined might empower people to acknowledge their embodied relations with other life forms, and that our fates are tied together. None of these initiatives involve a rejection of scientific findings, however – whether about the state of the land, or rivers, or the ocean. Just as Tupaia tried to share his understandings of the Pacific with James Cook and his Royal Society companions, while working with them to pilot the *Endeavour* and to draft hybrid charts of the Leeward

---

[34]Ngata (1928); qMs-1587, Alexander Turnbull Library, Wellington; Salmond (2019a).

Islands and the eastern Pacific, the Whanganui people are working with fresh-water scientists, and the new generation of Pacific voyagers with astronomers to explore and adapt ancestral navigational techniques,[35] and with marine scientists to to try and discern what is happening to rivers and the sea, and how best to return them to a state of ora.

In Aotearoa New Zealand, at least, where non-Māori as well as Māori are increasingly influenced by Māori values and conceptions, ideas of 'kai-tiakitanga' (taking care, guardianship) and 'ora' are being deployed to restore life to ecosystems. Where contemporary capitalism falls short, relational Māori phil-osophies may help to generate alternative ontological framings to guide everyday existence. As Geoffrey Lloyd has argued, modernity and Western science do not have a monopoly on truth or wisdom, and there is much to learn from other ways of living, and other understandings of what is real (2019, 41). As Merimeri Penfold once said, when understood as an equality, differences between groups (whether of humans or non-humans) are mutually defined, and insights are enriched by reciprocal exchanges:

He iwi kē, he iwi kē

Titiro atu, titiro mai

One strange people, and another

Exchanging perspectives with each other.

Disclosure statement

No potential conflict of interest was reported by the author(s).

References

Bambridge, Tamatoa. 2017. "Issues About Marine Tenure in French Polynesia." In The Sea Within – Marine Tenure and Cosmopolitical Debates, edited by Helene Artaud and Alexandrre Surralles, 81–96. Peru: Tarea Asociacion Grafica Educativa.

Beaglehole, John Cawte, ed. 1955. The Journals of Captain James Cook on His Voyages of Discovery. Volume I: The Voyage of the Endeavour 1769–1771. London: Hakluyt Society.

Beaglehole, John Cawte. 1962. The Endeavour Journal of Joseph Banks 1769–1771. Sydney: Trustees of the Public Library of New South Wales in association with Angus and Robertson.

Corney, B. G. 1919. The Quest and Occupation of Tahiti by Emissaries of Spain During the Years 1772–1776. Vol. 1. London: Hakluyt Society.

Crowe, Andrew. 2018. Pathway of the Birds: The Voyaging Achievements of Maori and Their Polynesian Ancestors. Hobsonville, Auckland: David Bateman Ltd.

---

[35]For instance, Nainoa Thompson's adaptation of the ancestral Hawai'ian star compass, which is now widely used by contemporary Pacific navigators, and Claude Teriierooiterai's work on Tahitian navigational methods.

Di Piazza, Anne. 2010. "A Reconstruction of a Tahitian Star Compass Based on Tupaia's 'Chart for the Society Islands with Otaheite in the Center.'" Journal of the Polynesian Society 119 (4): 377–392.

Di Piazza, Anne, and Erik Pearthree. 2007. "A New Reading of Tupaia's Chart." Journal of the Polynesian Society 116 (3): 321–340.

Eckstein, Lars, and Anja Schwarz. 2019. "The Making of Tupaia's Map: A Story of the Extent and Mastery of Polynesian Navigation, Competing Systems of Wayfinding on James Cook's Endeavour, and the Invention of an Ingenious Cartographic System." The Journal of Pacific History 54 (1): 1–95. doi:10.1080/00223344.2018.1512369.

Finney, Ben. 1986. "Re-Learning a Vanishing Art." Journal of the Polynesian Society 95 (1): 41–90.

Finney, Ben. 1998. "Traditional Navigation and Nautical Cartography in Oceania." In Cartography in the Traditional African, American, Arctic, Australian, and Pacific Societies, edited by David Woodward and G Malcolm Lewis, The History of Cartography 3.2: 443–492. Chicago: University of Chicago Press.

Finney, Ben. 1999. "The Sin at Awarua." The Contemporary Pacific 11 (1): 1–33.

Finney, Ben. 2003. Sailing in the Wake of the Ancestors: Reviving Polynesian Voyaging. Honolulu: Bishop Museum Press.

Forster, Johann Reinhold. 1778. Observations Made During a Voyage Round the World: on Physical Geography, Natural History, and Ethic Philosophy. London: printed for G. Robinson.

George, Marianne (Mimi). 2012. "Polynesian Navigation and Te Lapa – 'The Flashing.'" Time and Mind 5 (2): 140.

George, Marianne (Mimi). 2018. "Experiencing Mana as Ancestral Wind Work." Time and Mind 11 (4): 385–407.

Henry, Teuira. 1907. "Tahitian Astronomy: Birth of the Heavenly Bodies." Journal of the Polynesian Society 16 (2): 101–104.

Henry, Teuira. 1928. Ancient Tahiti. Honolulu: Bishop Museum. 355.

Lewis, David. 1972. We, the Navigators, The Ancient Art of Land-Finding in the Pacific. Wellington: Reed.

Lloyd, Geoffrey E. R. 2019. "The Clash of Ontologies and the Problem of Translation and Mutual Intelligibility." Science in the Forest, Science in the Past. Special Issue. HAU: Journal of Ethnographic Theory 9 (1): 36–43.

Low, Sam. 2013. Hawaiki Rising: Hōkūle'a, Nainoa Thompson, and the Hawaiian Renaissance. Honolulu: Island Heritage.

Matamua, Rangi. 2017. Matariki: The Star of the Year. Wellington: Huia Books.

Ngata, Apirana. 1928. "The Genealogical Method as Applied to the Early History of New Zealand." Paper presented to the Wellington branch of the Wellington Historical Association.

Ngata, Apirana. 2019. "The Terminology of Whakapapa. Intro. Wayne Ngata." Journal of the Polynesian Society 128 (1): 19–41.

Noyes, Martha. 2018a. "The Celestial Roots of Mana. Time and Mind: The Journal of Archaeology." Consciousness and Culture 11 (4): 371–384.

Noyes, Martha. 2018b. "Solar Nadirs in Pre-Contact Hawaiian Cultural Astronomy." Journal of Skyscape Archaeology 4/2: 201–228.

Oliver, Douglas. 1974. Ancient Tahitian Society II. Honolulu: University Press of Hawai'i.

Parsons, Harriet. 2015. "British–Tahitian Collaborative Drawing Strategies on Cook's Endeavour Voyage." In Indigenous Intermediaries: New Perspectives on Exploration Archives, edited by Shino Konishi, Maria Nugent, and Tiffany Shellam, 147–167. Canberra: ANU Press.

Pyrek, Cathleeen Conboy. 2011. "The Vaeakau-Taumako Wind Compass: a Cognitive Construct for Navigation in the Pacific." MA thesis, Kent State University, 84.

Sahlins, Marshall. 1985. "Hierarchy and Humanity in Polynesia." In Transformations of Polynesian Culture, edited by A. Hooper and J. Huntsman, 195. Auckland: Polynesian Society.

Salmond, Anne. 2003. The Trial of the Cannibal Dog: Captain Cook in the South Seas. London: Penugin.

Salmond, Anne. 2005. "Their Body is Different, Our Body is Different: European and Tahitian Navigators in the Eighteenth Century." History and Anthropology 16 (2): 167–186.

Salmond, Anne. 2006. "Two Worlds." In Vaka Moana: Voyages of the Ancestors, edited by K. R. Howe, 246–269. Auckland: David Bateman and Auckland Museum.

Salmond, Anne. 2008. "Voyaging Exchanges: Tahitian Pilots and European Navigators." In Canoes of the Grand Ocean, edited by Anne Di Piazza and Erik Pearthree, 23–48. Oxford: BAR International Series.

Salmond, Anne. 2009. Aphrodite's Island: The European Discovery of Tahiti. Auckland: Penguin / Viking.

Salmond, Anne. 2012. "Tupaia, the Navigator Priest." In Tangata o le Moana: The Story of Pacific People in New Zealand, edited by Sean Mallon, 56–75. Wellington: Te Papa Press.

Salmond, Anne. 2015–2016. Tupaia, a Documentary about Tupaia with Michael Tuffery and Lala Rolls.

Salmond, Anne. 2015a. "The Fountain of Fish: Ontological Collisions at Sea." In Patterns of Commoning, edited by Silke Helfrich, and David Bollier, 309–329. Amherst, MA: Off the Common Books.

Salmond, Anne. 2015b. "Introduction." In Lisa Reihana: In Pursuit of Venus (Infected), 1–3. Auckland: Auckland Art Gallery.

Salmond, Anne. 2017a. "Fountain of Fish: Moana/Sea." In Tears of Rangi: Experiments Across Worlds, 351–377. chap. 11 Auckland: Auckland University Press.

Salmond, Anne. 2017b. "Voyaging Worlds." In Lisa Reihana Emissaries, edited by Claire McIntosh, 42–65. Venice: La Biennale di Venezia.

Salmond, Anne. 2018a. "Reimagining the Ocean." In Oceania, edited by Nicholas Thomas, 42–55. UK: Royal Academy of Arts.

Salmond, Anne. 2018b. "Afterword. Think Like a Fish: Pacific Philosophies and Climate Change." In Pacific Climate Cultures: Living Climate Change in Oceania, edited by Tony Crook and Peter Rudiak-Gould, 155–159. Warsaw: De Gruyter. https://research-reposit ory.st-andrews.ac.uk/bitstream/handle/10023/16202/Crook_2018_Pacific_Climate_Cult ures_CC.pdf (2/10/20).

Salmond, Amiria. 2019a. "Comparing Relations: Whakapapa and Genealogical Method." Journal of the Polynesian Society 128 (1): 107–129.

Salmond, Anne. 2019b. "Hidden Hazards: Reconstructing Tupaia's Chart. Forum on Tupaia's Chart." Journal of Pacific History 54 (4): 534–537.

Salmond, Anne, and Amiria Salmond. 2010. "Artefacts of Encounter. History and Human Nature." Interdisciplinary Science Reviews, edited by Brad Inwood and Willard McCarty, 35 (3-4): 302–317.

Teiwa, Teresia. 2015. "Manukau NZ: Tagata Pasika and Manukau Institute of Technology Pacific Education." https://www.youtube.com/watch?v=lipupbIZb6U (1/10/20).

Teriierooiterai, Claude. 2013. "Mythes, astronomie, découpage du temps et navigation traditionnelle: l'héritage océanien contenu dans les mots de la langue tahitienne." PhD thesis, l'Université de la Polynésie française.

Thompson, Christina. 2019. Sea People: The Quest to Understand Who Settled the Islands of the Remote Pacific. London: William Collins.

Turnbull, David. 1998. "Cook and Tupaia, a Tale of Cartographic 'Méconnaissance'." In Science and Exploration in the Pacific: European Voyages to the Southern Oceans in the 18th Century, edited by M. Lincoln, 117–131. Woodbridge, Suffolk: Boydell Press.

Tyerman, D. 1831. Journal of Voyages and Travels by the Rev. Daniel Tyerman and George Bennet Esq. . London: Frederick Westley and A.H. Davis.

Wi Repa, Tutere. 1923. "Maoris of East Coast: Research by State Ethnological Party: Visit of Distinguished Scientists." Gisborne Times. 12 April 1923. Quoted in Salmond and Lyth.

Counting generation(s)

Marilyn Strathern

ABSTRACT
Papua New Guinea is known for its many body-based, counting systems and for people's passion for enumeration. Neither seems to require a unified concept of one. Rather the 'instability' (after Vilaça) of one turns out to be germane to the very facility to get from one number to another. The movement that is counting echoes replication and generation in other registers, as in the life of food plants propagated vegetatively, and in the life of people with its marked sense of displacement and replacement. Insofar as counting seemingly embellishes such processes, the latter, in turn, offer insight into practices of computation.

Bamboozled by numbers: this is how Ingold (2019, 667) describes the penchant for 'adding things up' prevailing among 'those of us educated into the ways of modern science'. But to add things up, he says, they first have to be broken off from the ebbs and flows of life and the processes that give rise to them. For life is ever becoming, forming and dissolving, folding into its future; thus, in the unison of song, you may, 'through differential attention, be able to tell one voice from another, to split them along the grain of their becoming. But you cannot count them up' (2019, 668). Enumeration appears inimical to life.[1] Ingold's criticism that, rather than begin with everything, 'our' (argumentative) inclination is to conclude with everything (hence the lure of addition and accumulation), resonates with much that is known of Euro-American scholarship. This makes interesting those cosmologies that intertwine counting and regeneration: here number may have much to do with the perpetuation of 'life'. The morphing of beings into one another, the impossibility of extracting singularities from continuous process, generations repeating themselves – these may be imagined with apparent elements of counting systems that also enable feats of arithmetic. If Melanesians of former times did indeed bamboozle themselves with number, the dazzle was not divorced from life's unfolding.

Tracing what kind of life this is calls on a diversity of materials that will, I hope, make the point indirectly. Asking what kind of number demands more

---

[1] It is life with a generative cast to it, perpetuating a sustainable planetary future, which he has in mind (2019, 670–671).

direct treatment. For we at once run into the issue that Vilaça (this volume) adumbrates: while counting may start with one, in old Melanesia[2] it frequently turns out to be as unstable as its Wari' counterpart, and indeed we shall see that 'one' is far from the obvious starting point it might seem.

If such a number cannot be held steady, then how can entities be stacked up, proceed from one to another, or indeed added together?[3] Like Borges's pebbles, how can things cohere long enough to be enumerated before they start changing? How do people generate items to count at all? Answers of a sort may lie precisely in what gives the impression of instability. The article begins with some of the ways in which entities are conceived; these bring it to certain regenerative practices, which, in turn, lay grounds for describing Melanesian counting systems. Ingold is right to be cautious about where one begins and ends, and in a manner of speaking I too wish to begin with 'everything'.

Varieties of instability

More than 50 years ago, M. Panoff elaborated on the concept of the 'double self' that Maenge, then some 5000 people in New Britain (Papua New Guinea), call *kanu* or shadow. It conveys a notion of likeness, such as a noise like crying, or of prefiguration, one event announcing another; 'an element considered as a specimen of a whole set' (1968, 276). *Kanu* underlies appearances. Felled timber left in the open and subsequently swollen with rain is said to have recovered the *kanu* it had when a living tree; the emerging form of a canoe is the craftsman's *kanu* embodied in wood. *Kanu*, M. Panoff suggests, is the self of an object, and 'in the human realm is to be regarded as the self of the person' (1968, 277). While the living person is housed within an outer container, a reference mainly to skin and bone (like walls, roofing), it is the body's inner substance that is suffused with the self. Fluid-like, *kanu* permeates this substance, yet escapes it from time to time, as when a craftsman's *kanu* enters the canoe he is making.

There are two doubles here. Just as the physical body comprises outer container and inner substance, so too *kanu* is divisible. The outer self is an unseen form that fits over the body, not adding thickness but controlling its shape; viscous, adhering to other things, it is regarded as 'dirty'. By contrast, the

---

[2]The phrase 'old Melanesia' was originally Alfred Gell's. The present synthetic account dwells on a few ethnographically located instances from among the diversity of Melanesian practices; time horizons are also diverse, but relate largely to pre-Pentecostalized societies. (Melanesian diversity is to be taken seriously; thus while the people of Ponam in Manus Province [Carrier 1981] show huge interest in counting, and can extrapolate to hundreds and thousands, they neither count people nor relate numbers to body parts.)

[3]Myhre [see Acknowledgements] queries where instability lies, not in the number (he suggests) but in the entities to which it is applied. In these contexts, the aesthetic – the recognition of a particular form – seems stable. I keep with the original formulation, however, as a provocation to the vernacular English-language assumption that one implies a singular unit with or on which operations can be performed. As we shall see in certain instances from Melanesia, as a relational element or effect of enumeration 'one' cannot possibly be such a unit, and indeed it is clear to the Papua New Guinean Paiela that a one by itself is uncountable (Biersack 1982).

'clean' inner self, sometimes caught in a reflection, controls a person's health and beauty through the liveliness of blood or breath. Either self may wander away in sleep or in sickness. Should the physical body lose both selves for more than a short period of time, the being becomes lifeless. Conversely, inner and outer *kanu* can only be held together by that which they keep alive, namely the inner body with its outer protective housing. 'In fact, it would be no exaggeration to say that the whole life cycle of the Maenge [person] is spent in efforts of recovering or keeping both ... *kanu* [together]' (1968, 279).

It is in striving to stay alive, then, that persons work to keep all their aspects together, yet not only leave bits of their external self in the world around but also through their activity breathe into that world something of the inner self. Keeping everything together suggests we could singularize the person as a number, while noting that being one signals a veritable achievement.[4] For 'one' comes as an entity at once divisible and multipliable. Thus are living beings divided from the dead and made manifest by them. That one can say '5,000 Maenge' is actually cause for remark. Of course, the doubly doubled self-and-body need be no more problematic for the arithmetic required than the Euro-American assumption that an individual plays many roles – it is not the number of roles that gets into a census. That aside, Maenge calculations would seem oriented elsewhere. Scepticism is directed at official census takers who focus on living clanspersons and neglect the selves of the dead (1968, 286): the officials have obviously mistaken the purpose of counting. So what does it mean to count people – and, as we shall see in systems involving body parts, count *with* people?[5]

That magnitude is significant holds for Melanesia more widely. Where the instability of Wari' number accompanies lack of interest in enumeration, Melanesians have been known to go to the other extreme in producing 'large' numbers. Indeed, for enthusiasts of wealth transactions, whether over the life-cycle, in war indemnity or for the sake of exchange itself, counting is indispensable. Computations range from the massive amounts of yams – several thousand baskets – competitively assembled for the Paramount Chief of Omarakana on the Trobriand Islands (early twentieth century); to the elaborate (mid-century) denominations and conversions of Kapauku shell money, to (late century) Papua New Guineans calculating royalties on mining concessions commensurate with an enlarged sense of self-worth. Pospisil ([1963] 1978, 94), for instance, wrote that Kapauku men, using 'a decimal system that stops at 60 and starts over again, having as higher units 600 and 3,600, ... count their wives, children, days, visitors at feasts and, of course, their shell

---

[4]Telban (1998, 227) uses the vocabulary of achievement in just such a context.
[5]Myhre's stimulating reading makes me add that this is a rhetorical staging post in the argument. The English 'body parts' is a bit misleading; the point is that, whether or not the image of a whole person is an explicit part of enumeration, people count on themselves.

and glass bead money'. To show appreciation, the ethnographer might invite Kapauku associates to count his store of trade beads, a satisfying pursuit that could last hours.[6]

High-order enumeration may be competitive, and open to dispute thereby. Yet disagreements can occur with the smallest of numbers. As reported of the 1980s, married couples in the Murik Lakes area of Papua New Guinea sometimes diverge in reckoning how many children they have (Lipset and Stritecky 1994).[7] Murik men readily include children who are adopted (in and out), while women recall deceased offspring whom their husbands omit. Adoption supplements men's socially procreative force, since their larger nurturing efforts include helping others to have children; conversely, they deny subverting that very force through neglect of certain observances (infant mortality incriminates the father). The public image of maternity is of generosity and indulgence, child-bearing enabling women to show their care, and although they sometimes mention children adopted-in will ignore those adopted-out. This divergence of parental influence echoes the moral tension between what Lipset and Stritecky (1994, 3–4) call inner and outer body: the 'negative valuation of mystical fluids, spirits, and processes that take place within the body and the positive valuation of activities undertaken by and ceremonial appearances of the exterior body'. One effect is difference in the relational targets of harm and protection: men can injure their own children, where women can be lethal to adults; men nurture ('mother', so the idiom goes) adults, just as women nurture children.

Murik children are thus divided by the perspective of male or female parent, now offspring of a mother, now offspring of a father. Similar calculations are explicitly embodied in those kinship structures – not true of Murik – with a 'descent group' (matrilineal/patrilineal) cast. Persons are divided by their origins in the sense of having to manage their issue from clans on both mother's and father's side. Each side makes its own claims, with blessings and curses to bestow, in relation both to one another and to the next generation. The two kinds of kinship bodies must be managed in tandem. Considering the often elaborate prestations that pass between kin related through marriage – bridewealth, birth ceremonies, funeral payments – we can borrow from Maenge and say that in these regimes the whole of a person's transactional life is spent in the effort of keeping paternal and maternal sides together. What is deliberately held together, also divides into two. Through such prestations, each side acts momentarily in respect of a specific other; conversely, the

---

[6]This famous passage illustrates Kapauku obsession in the matter. The 'fortunate individual often squatted over my boxes for as many as four hours. At the end of his counting he would report my wealth with a victorious smile: 'You have 6722 beads in your boxes. That means you have spent 623 beads since Gubeeni counted your money three days ago. I would suggest that you order more beads in about thirty days' (Pospisil 1963, 94). Counting seems to have been something of a local specialization, Kapauku enthusiasm far exceeding that of their neighbours.

[7]In one instance they render the discrepancy numerically, but it is not clear if their general references to the 'number of children' are based on Murik enumeration. It should be added that disagreeing parents were in the minority.

division retains them as a pair. A single parent is created as a 'one' that is half of two.

We saw that, as with the diverse locations where a Maenge person's *kanu* lodges, the objects of Murik nurture are extended to increase the parental capacity mothers or fathers claim: living proof of its magnitude lies in how many children are separately named. Under other circumstances, 'one' expresses a magnitude that can be given no higher value. Where parental idiom is used, an aura of omnipotence may single out one of the two parents thus.

From men's perspective, something like this is evident in old Hagen (and see Merlan and Rumsey 1991).[8] 'One father', or more generically 'one man', becomes an encompassing reference point for a clan, or better put, a condensed figure of it, a kind of end point, with a visibility that big men especially work to engineer. On occasions of public display, the clan acts as one man with respect to diverse partners and onlookers, momentarily submerging internal differentiations and the equally diverse loyalties of fractious and independent-minded clansmen: the magnitude of a big man's achievement rests in how he creates one out of many. The crucial maternal–affinal ties so important from the viewpoint of a person's life cycle are occluded by the clan's capacity to regenerate itself. (Patrilineally-speaking, son follows father.)[9] Thus a clan, engaged (say) in arranging war compensation, marks itself off from specific exchange partners of the moment – frequently maternal and affinal kin – and presents a single, unified line to a heterogeneous audience. Men astound themselves with its coherence: we have made ourselves one man! They then turn from dancing as a triumphant 'one line' of performers to attend to their exchange partners, each of whom now appears as the counterpart or pair of each of them.

Ones emerge in diverse situations, then, and would indeed appear unstable if an observer tried to add them up. While one can be a stopping point (not so much greater than a multitude, but rather a whole that is *other* than a multitude), it is also impossible to think of one without thinking of (a divided) two, and keeping a pair in tandem seems to require as much effort as keeping a one together. There seems no simple basis for assuming containment or identity.

Replacement/regeneration

The effort that some Melanesians put into marshalling their multiple selves or acting like one person implies a being that endures. Collectively, such an entity

---

[8]In the Papua New Guinea Highlands. I began work in the Mt Hagen area in 1964 (initial accounts include M. Strathern 1972, A. Strathern 1972); the present article also draws on the work of missionary-anthropologist Strauss ([1962] 1990). Merlan and Rumsey (1991, e.g. 95–98), describing for Hageners' Ku Waru neighbours the grammar of person/number in terms of 'segmentary person', introduce significant qualifications on how the apparent totalisation of 'one' may be understood.
[9]And vice versa in those regimes where marriage rules have a son repeating the direction of a grandfather's marriage, thus by assimilation becoming father to his own father.

may comprise not just all who have been but all who are to come (e.g. Mimica 1988); at the same time, new generations appear as transformations of previous ones, a matter more of replacement than succession. When a (grand)father reappears in his (grand)son, the latter is taking the former's place, a kind of self-exemplification. Indeed, there is a profound sense in which replacement has no time depth to it, for what is replaced from the past is evinced within the replacement, while the latter's own destiny is to be replaced in the future.[10] We can add a self-perpetuating 'one', of a repeatable or recursive kind, to the repertoire.

Vernacular formulations of self-perpetuation abound, rather closer to cloning than to the lineality that inspired anthropologists to write of descent groups. It is true that the ethnographer of a yam growing area (Coupaye 2013, 290) is writing of such groupings when he remarks how clans are regarded as (yam) tubers emerging from one another, and thus 'as variations of one another': some 'are older than others, some came first, others followed … But all are related … through the first tubers. So are banana[s] … , so are people, so are places … They emerge out of each other, like Russian dolls'. Yet the group-like appearance of a clan is as much outcome as precondition. In any event, the replaceable social entity could as well be a locality or a set of siblings. This last is the case for the taro-cultivating Kaulong, who like Maenge live in New Britain. They prompted their ethnographer to comment on the model of parthenogenesis offered by vegetative propagation. With reference to the 1960s, Goodale says: 'men's model of social reproduction is one based on the seemingly asexual reproduction of tropical plant-life where new generations are cloned from the old' (1995, 158). The next generation is like a tuber or corm springing anew from the soil as the present generation sprang. This does not deprive it of particularity. On the contrary, Coupaye's term 'variation' is very apt: it captures the generalized possibility of entities existing as varieties of one another, such that a particular time (or place) produces a variant of a being already in existence.

The apparent timelessness of replacement does not, then, obliterate particularity. Rather, 'one' entity may be manifest in numerous versions.[11] As reported from so much of Melanesia (Rumsey 2000; Moutu 2013), a man will talk of what has happened to him and mean some ancestor, another manifestation of himself. Horticulture is an area that generates enthusiasm about variation, and people keep track of where plants have been, actively seeking out varieties

---

[10]A concern with 'replacement', for which vernacular terms exist, is found in the idea that when people die their place will be taken by specific others (e.g. see Scaglion 1999, 221; Goodale 1995, 158). It may be elaborated as a three-generation process (grandparents and grandchildren being identified with one another) or across adjacent generations between persons paired by some specific compatibility.

[11]I am cautious about deploying the concept of identity. Coupaye (2013: 290) refers to the imagined clan entity as a prototype. Salutary would be the linguistic perspectives offered by both Rumsey (2000) and Myhre (2016), and the significance of bringing different states into being through naming. (It is when a 'variant' is spoken of or named – given a presence in language – that it is what Euro-Americans would call 'identified' with a recursive ancestral being.) Such significance would be consonant with that I eventually give to Melanesian number.

of cultivars (for example, F. Panoff [1970] 2018). So, Kaulong gardeners pay attention to the social sources of taro plants, sometimes to reinforce the effectiveness of a cloning-like replacement,[12] sometimes trading or exchanging across diverse networks the taro stalk that will be a source of new plants, so their taro are grown in other places, by other people, just as some of theirs have come from elsewhere. This sense of variation valorizes outside sources. Regeneration is not channelled in one direction only; it spreads over a range of relations.

Indeed, if we stay with the plant analogies Melanesians are so fond of, there is more to replacement than repeated (re)appearance or the creation of like from like. Those wider networks come into it, but the crucial issue is that the appearance (of a previous generation, say) is also an event.

Growing yams send vines above ground that causes nutrients to be stored underground in the swelling tuber, the vines dying back before the yam is harvested; taro grow thick stalks and leaves above ground, and the portion of corm to be re-planted will have some stalk attached. In either case, once out of the ground, the tuber or corm ceases to grow, but when people *slice* a piece from it – or in the case of some yam varieties whole tubers are separated from the rest – a new point of growth is created. Crook (2007, 95) describes the constant unfolding of a taro plant. Each leaf dies after it yields its successor, until the dramatic moment of harvest when growth is altogether stopped and the taro to be eaten is severed from the stalk to be planted. It is crucial that adhering to the stalk is that fragment of the old corm; planted, it nourishes the new one growing in its stead. Capacity for future growth is thus held within it. Here we find that the concept of cloning, in the English sense of serial reproduction without change, does not quite capture the process. A new yam tuber/taro corm might 'timelessly' repeat what it replaces, but – and what follows is my commentary – that replacement is not regarded as automatic. Cultivation practices require a crucial intervention: the plant has to be planted. And insofar as the replacement has thus to be cultivated into being, it also exists in a temporal relation to its predecessor. Past is divided from present. The relation is binary; there must be specific substitution or reappearance of one thing by another. Among people, the future returns of past generations may be written into marriage rules that repeat (in the same or other alignment) the direction of earlier alliances. Yet, such a recursive 'one' has always to be *made* anew,[13] just as there will be no yam or taro without the cultivator's action.

An existential issue yam and taro cultivators might recognize is how to produce the next generation of parents. As plant material can only be generated from plant material, so too persons: their unfolding has to be kept going. Yet, in

---

[12]'When someone said, "This is my grandfather's taro," it was clear that it was not only the same variety as the grandfather had planted, but it was considered to be the identical plant' (Goodale 1995, 77).

[13]Effort ('work') becomes manifest in people's resort to 'magic' or 'ritual', considered crucial to generativity and growth (e.g. F. Panoff 1969, 21).

the case of persons, how can parents produce parents? The answer is in that intervening, binary step. Parents make children (not parents) for it is, in turn, children who are made parents – in old Melanesia through bridewealth, initiation and similar cultivation techniques (my phrase). Division, too, takes effort.

It is there in the contrast between the regenerative possibilities of the parental parts of food plants, with future growth within them, and the already grown tuber or corm, now inert. The same plant might reappear over and again, but only a terminal cut between the generations can create the new one that will keep life flowing. Partitioning is an event. In lieu of allowing spontaneous growth from rotting material, people precipitately divide the living plant, stopping its own growth, taking what they want for food and replanting or discarding the rest. While growing in the ground, the yam or taro is in many places idiomatically called a 'child'; sometimes the harvested child is spoken of as an ancestral spirit (grandparent) who has emerged again. The severed fragment retained as generative, pushed back underground, shrivels as its own new child grows, first nourishing the young shoots and then dying.

Thus too the propagation of persons: a binary divide between adjacent generations (parent and child, in this sense unlike each other) also supports relations across alternate generations (child with parent's parent, like with like). Life-giving parents are moving towards death; death-transformed children spring up with their life, and procreative future, ahead. Consider how Kaulong asexual cross-siblings – brothers and sisters – replace themselves as a sibling set. Since siblings cannot in actuality, as a plant does, regenerate through cloning (Goodale's phrasing: 1995, 180), they must reproduce sexually. People marry outsiders in order to have children who are their own personal replacements, it being the sibling set of each spouse that is regarded as socially regenerated. Goodale argues that the perceived self-reproduction of the sibling set is a denial that sexual activity is necessary for future life, even though (we may add) death is. Kaulong are explicit about the parents' fate: they die in creating a new sibling set in place of the old. So while men and women alike desire replacements, anticipating the replacement anticipates, precisely, their replacement.[14] The disappearance or displacement of the predecessor guarantees the appearance of the replacement.

Drawing from Moutu's invocation of necessity (see below), we might say there is a cosmic necessity to keep death going as much as life. Maenge make something of that necessity evident in terms of the mutual implication of people and spirits and their accompanying benefits and dangers. Recall that the person has both 'self' and 'body', each doubled in its capacities for well-being or for harm, and that bodies and selves have to be held together. Their

---

[14]Young man: 'When I am old and ready to die, I will get married and find a replacement for myself who will bury me' (Goodale 1995, 133).

other ethnographer, F. Panoff (1970, [1970] 2018), translates the 'self', *kanu*, as soul. These cultivators, she says, ascribe to taro the same theory of soul that they entertain of themselves. Taro souls are under the control of various deities or (spirit) masters who have to be coaxed into letting the souls stay attached to the growing plants. Like people's, taro souls may get up and wander away, and are quick to take offence if not properly cared for. It is its soul that makes the corm heavy and nutritious, and effort has to be exerted to keep souls tethered to the plants or the harvest will be worthless. Given that these souls are refractions of a generic taro soul, dedicated action is necessary to attract as much soul as possible into the growing corm. Echoing Wari' ideas about the need for the continuous accretion of potency throughout life, the generic soul has to be replenished, but properly replenished exists in perpetuity, just as might be said of the perpetuity of Maenge matrilineal kin groups. As is true of peoples elsewhere in Melanesia who do not necessarily share Maenge notions of plant souls, any particular being is part of a generic life-force. What is interesting is the Maenge work of division.

If body is that which is animated, reciprocally it is also that on which the soul's own regeneration depends. Maenge make explicit the soul's need for a conduit, what passes though people's bodies being in counterpoint to what passes through spirit bodies. Mythically imagined, food is the excrement of spirit beings, often depicted as pythons, while people's bodies are equally con-duits for the growth and continuity of taro soul. Taro soul (the nutritious part) passed through the bodies of spirits is thus called 'snake excrement'. And taro soul, as it passes through the bodies of people, is, in turn, augmented by diges-tion.[15] In short, soul requires constant replenishment in the things it animates. We have already encountered its 'leaky' aspect. F. Panoff remarks that while the greater part of the taro soul – said to be its liquid – is absorbed by the consumer, some of it remains stuck to faeces, faeces being the unassimilated fibres rotted down in the belly (1970, 243).[16] Such virtual faeces, and the fertility they imply, are considered a prime matrilineage resource, valuables such as adze blades and shell money also being called 'snake excrement'. One consequence is that people must scrupulously avoid external contact with people's faeces (which provoke feelings of revulsion), exactly in order to perpetuate the cycling of life force impelled by the food/spirit excrement coursing internally through themselves.

What nourishes some poisons others, depending on whether a person is spirit or non-spirit. But neither state is remote from the other: life and death thrive in mutual dependence. Moreover, the various divisions here take the

---

[15]Primeval man, the Maenge story goes, had an open head through which food would just fall through his body because he had no digestive system, that is, a proper passage from mouth to anus (1970, 244).

[16]'Rotting matter is thus the substance of food – as well as its source and its end' (1970, 250); ignoring the inter-vening conversion of faeces into food in effect allows a parthenogenetic reading to the endless cycling of waste. The 'rotten' includes corpses.

form of events insofar as they are embodied in observances and taboos that require people to differentiate certain activities from one another, including taking care of their power to contaminate others. If Maenge have to cajole spirits into helping them, they also have to be mindful not to interfere in the complementary cycling of life and death through spirit and non-spirit beings and thereby compromise the difference between them. If, insofar as it is achieved, 'one' is an event of sorts, the same may be said of sustaining crucial divisions into 'two'.

Kaulong desire for (life through) replacement has its other side in people's terror of dying. This makes the latter no less a necessity. Moutu (2013, 147) applies the concept of necessity from Western philosophy to his study of the Papua New Guinean Iatmul. The persistent pairing of entities in the Iatmul world is epitomized in the elder brother-younger brother duo, in ultimate terms paired as life and death. Death overtakes life, and life overtakes death; each is a version of the other. This conception organizes certain dispositions (such as slow and fast action) and, applying to men and women alike, organizes reference to all kinds of relations across the generations or between the ritual moieties that initiate one another's youth. Grandfather reappearing in grandson is designated elder brother to younger brother. Brotherhood is thus, of neces- sity, oriented towards a certain kind of emerging, 'one [brother] forever becom- ing the other brother' (2013, 202). The process is simultaneously generative, dynamic and event-ful. It brings us to a juncture where we might begin to understand counting.

Imitating a Melanesian proclivity, drawing on plants to talk about people and vice versa, reinforces the way in which ones appear out of what Euro-Ameri- cans would call organically related processes of growth or regeneration, includ- ing the self-perpetuating one-ness of an entity whose diverse variations do not affect its already existing character. As we have seen, beings are often posed as versions of each other, whether multiple (the many persons of a clan) or dual (inner as opposed to outer soul). It also reinforces the point that attending to the Melanesian one summons the inevitable cutting or division that intervenes to make propagation other than a process of cloning, and yields the kind of 'one' that is also a 'two'. A multitude (see p. 290, line 189) disappears when many ones each appear as a refraction or version of an ultimate one that is them all, while the division of ones into two creates a pair: either state of affairs may be closer to a logical-virtual starting point for counting than trying to isolate an original singularity.[17]

---

[17]Vilaça (2019, 50–51), who also instances several Melanesian systems of body-counting, discusses the relative valorization of one and two, going back to Levy-Bruhl's observation about many numerical systems beginning with two. M. Panoff reminds us of nineteenth-century interest in Melanesian enumeration, and early twentieth- century assumptions about peoples who tend to count in twos, as well as recording the Maenge associate who explained, 'The white man counts one, one, one, but the way of our ancestors was to count by twos' (1970, 362). Panoff's detailed analysis of the three cyclic patterns that Maenge deployed in counting leads to an understand- ing of addition, subtraction and multiplication. However, Biersack's (1982) account of Paiela enumeration in

I have briefly drawn attention to moments of extension or leakiness, which work against the containment of what Euro-Americans would consider identity. Then there are those transfers and exchanges across clan groups or ritual moieties, or between spirits and non-spirits, through which otherwise differentiated persons participate in one another's life-bestowing/death-dealing powers. And in the movement generated by the event that is counting, we glimpse the pull of one thing by another. A famous rendering of one of Papua New Guinea's famous counting systems puts this into the language of number.

Enumeration

The Iqwaye, Mimica (1988, 13) remarks, count all manner of things – dogs, taro, days, people.[18] Before a raid, assembled fighters would stand in line and a rope of small shells be measured against them, analogously one man, one shell, to determine whether they were enough to go ahead. The 'one-ness' of the body of men is exemplified in the tally as a magnitude. But here he makes a crucial difference between the single rope of shells and the possibility of enumerating the raiders. They have not been added together to arrive at a total; rather, their unity transforms many into one, and however many men are present the numbers themselves do not index the 'one' that is all of them. So what is it with number? After all, Mimica (1988, 19) emphasizes Iqwaye concern with quantitative exactitude, as in finance where 'they tend to reproduce meticulously the most minute quantities of things in numerical values'. If the rope tally is not itself a number-object, a person's digits are. With them, counting can be applied to almost anything, just as it could be said that 156 or 161 men made up one rope.

Mimica's language of parts and wholes is apposite when it comes to number counting. Iqwaye take digits as parts of a hand, and all someone's fingers and toes as parts of (composing) 'one' person, 20 – two sets of hands and feet – being the base of Iqwaye enumeration. Here wholes are, in effect, the outcomes of a movement from one place to the next. Thus 5 is expressed as 2 + 2 + 1, that is, 1 (whole) hand. Other computational possibilities follow. Six is one hand (5) + 1; 24 is 20 (2 hands + 2 feet or 1 person) + 2 + 2; 105 is 5 persons, each a one with all its 20 digits plus another 5, as

---

terms of the binary structure of communicational processes offers the most complete and elegant address to a relational logic where the minimal constituent of a pattern is not a unit but a pair. It is a paean to Paiela powers of abstraction. My own emphasis on attention to processual replacement as a starting point for (interpreting) counting actions – and Pickles [see Acknowledgement] remarks on this – points to such abstraction as immanent, so to speak, in (across) non-numerical fields.

[18]As do Maenge: 'In the old days, informants insisted, all things were counted – the booms of outrigger canoes as well as the pegs inserted into the float below, the sticks used in gardening magic as well as the leaves collected for divination purposes, and of course pigs and the shell rings recognized as ceremonial money' (M. Panoff 1970, 364). Coconuts were counted by fours, five such (20) forming a traditional trade unit; taro brought ceremonially from a garden were gathered in sets of 20.

though from another person. The plus signs are the ethnographer's. When they express their actions linguistically, and using just two number words, 'one' and 'two', Iqwaye speak as follows. On the left hand, beginning with the thumb, the first digit is 'one', the second 'two', the third 'two-one', the fourth 'two-two', the fifth being a 'hand' and a point of initial closure. When they proceed beyond the first hand, the term 'to-the-next' indicates a movement to the other hand (so 6 is 'hand to-the-next one'), just as involving the toes is '[two hands] down-to the foot one'.[19]

In the light of other materials from Melanesia, a couple of features are worth stressing. First, Mimica (1988, 51–52) notes that 'one hand' is a closed series. The primary seriation is of binary succession (1 – 2 – 1 – 2 – 1), not of ones alone (1 – 1 – 1). So 3 is not summatively 'one and one and one' but more like a located or (re)placed 'two-one', which carries some of the senses of an ordinal (the 'second' one). In other words, it is like the first 1 but – as with every generation or harvest – in a different place/time.[20] Second, the apparent sequence [the formulation is mine] by which digits are aggregated (into one) and disaggregated (into two), making one simultaneously half of two and the togetherness of two, echoes some of the processes of division and pairing at work elsewhere. Basically, each counting movement either adds one to a previous two or completes a previous one with a further one; a pair acts as one and can be completed in a similar way, by another one/pair.

It is not the singularity of the digit but the division or partibility of the body that is seized upon.

> The corporeal person intended as one but representing 20 persons or 400 digits is identical to each and all of them. It is identical to each of them as a whole, as a part, and by the number of its parts. That is, each person as a multiplicity of 20 is in itself *one* person, therefore a whole equal to one part (one finger) – it is totalized and thereby partialised. As such it is an exact replica of the oneness of 400 which itself is also a totalized and partialised whole … unified into a whole … rendered equivalent to its single part. (1988, 57 original emphasis)

Mimica warns against imagining this multiplication in terms of Western conceptions of number construction (1988, 40–41): the Iqwaye computation of 400 is simply 'as many persons as my [speaker's] digits', that is, each person having 20 digits and as many of them as the speaker's (20) digits. (400 becoming one person then affords a new base for higher evaluations.) All these persons, each being a set of 20 digits, are represented by the same body and digits that someone uses to count from 1 to 20.[21]

---

[19]Mimica uses the term 'leg', which includes foot. One foot is not just five toes but presupposes the two hands (10 fingers), which are laid on the toes as people count.
[20]Karine Chemla's pertinent observations at the symposium urged me to specify this operation.
[21]Mimica stresses that you don't count numbers, you count things, and Iqwaye things are the body and its divisibles. There is no exterior 'environment' of already discrete entities to which counting applies. Everything is within (the body/cosmos).

Enumeration was originally propagated by the mythical paternal figure, Omalyce, of whom all persons are parts. From a patriarchal point of view, the cosmic figure is equated with the totality of his progeny and thus with every single part of himself. The first set of men he made out of mud, 'Omalyce's replicas – as it were, his clones ... five individual wholes named after the parts of the whole' (1988, 83). The omnipotent father is both gentitor and progeny. But then again the fifth son, whom the creator figure produced out of himself, also turns into his double, a progeny-bearing woman. The one-two alternation of fingers is also that of male (oneness) and female (twoness).

> Symbolically, numbers are humans and they are, appropriately, generated like them ... . Through the first pair of the one and two, all other numbers are generated as their replicas, their multiple doubles. They are seriated as the pairs of complementary opposites, for only in that way they can reproduce ... [T]he two is the double of the one which through it ([the] two) doubles itself, begets another one, and so on. (1988, 92, 94)

The generativity of counting literally entails an act of generation.

We return to the question of spatio-temporal displacement. A repeated alternation, one into two and two into one again, terminates itself. So what is implied in termination? Vilaça (this volume) speaks of Wari' valorizing the two or the pair, but not passing from two to three or four, nor turning two into one. Incessant as Wari' are in their reproduction of pairs, distinction is always sustained, never leading to equivalence between the two parts of a pair. Wari' counting allows us to specify a rather different process when it comes to generalizing for Melanesia, and no less for Iqwaye. People seek their replacements (future life), with the replacement identical neither to themselves – singly or as a couple – nor to previous generations. The grandparental generation returns but in a 'new time' (M. Strathern 2019). In brief, the 'one' that is produced out of 'two' both *is* and *is not* the same as the one that produced the two in the first place: it is a new version – a para-equivalent – of that one.

Perhaps the ever-present transformation of ones and twos gives rise to something like the surprise of a harvested taro: what was divided springs up again as an undivided entity, at once a timeless replacement and ushering in a new time. Forward movement does not engage linear or evolutionary process but anticipates the next time that will replace (displace) the present. Perhaps some of the compulsion of forward movement is captured in Pickles's (2019, 55–56) account of the value of speed in card games, which is behind the present popularity of a markedly 'fast' game in Goroka town. (One pattern of cards is displaced by another, and the speed is in the turnover.) Apropos that movement more generally, 'one' cannot be a point of stability insofar as it is trying to be something other, as the Iatmul elder/younger brother pair tell us;[22] 'two' may be a generative

---

[22]This trying to be other is not true of all pairs (see footnote 25). Iatmul elder-younger has a specifically generative/generational ('cross-sex') inflection, as do ones and twos.

coupling, but cannot reproduce itself as a pair. That all said, the dynamic of disaggregation (procreative cutting) and aggregation (the inert pre-condition/ outcome or stopping point) can inform all manner of Melanesian being and doing without necessarily taking explicit numerical form. I infer that those diverse practices are not in themselves about counting, and do not yield number-objects. As an enactment of generative logic, numbers are a special obsession.

Yet all the multiplications and divisions that persons perform on and with themselves throw light on the very compulsion of numerals. 'Obsession' comes from Pickles'[23] further comment on how Melanesians show off their counting skills – it is one of the 'joys of life'! People frequently count when they want to display their (positive) efforts, as Hagen men take omens of future well-being before a display or as Iqwaye do before a raid. Indeed, for all that it is a tool or an operation that can be drawn into any manner of skilled application,[24] enumeration also entails an aesthetic that calls forth a response, as a form not itself embedded as such (in that form) in other activities but activating its own conventions. So while I have stressed unbroken continuity being perceived in tandem with entities severed from one another to effect replacement, whether in terms of body and soul, in a botanical or gender idiom, or as a matter of death as the prerequisite for life, counting at once springs from these modes of being/doing and is its own version of them. In this, counting is – to recall Ingold – just like singing. Song deploys a particular repertoire of actions, appearing as an event, even as it gives voice to events elsewhere in people's lives.

Envoi: the pull?

It was in hearing me give a Hagen version of hand-counting at the symposium that Blackwell (see this volume) drew another comparison with music, apropos the time being structured through the sequencing of digits. Entities   otherwise   seemingly   discrete   flow   into   one   another.   My

---

[23]My thanks for his illuminating commentary at the symposium. As he has elaborated (pers. comm., 2020), the joy is of bringing uncounted things into relationship through a person's organizational agency in counting (and see Biersack 1982, 814–815). Hence, he observes, people's fascination with the kind of sedimented, congealed, or manufactured consistencies that emerge out of a roiling, boiling soup of potential connection, and thus with games, rules, with money itself, with business and projects and self-fashioning to create something memorable.

[24]The PNG University of Goroka is well known for one of its educational specializations, the Glen Lean Ethnomathematic Center. Oriented to improving mathematics learning in schools, it has found huge interest among Papua New Guinean teachers for tools enabling them to teach the subject with materials that already make ('cultural') sense to pupils (e.g. Matang and Owens 2004). Such teachers are enthused by the idea that what appear aspects of modern or scientific mathematics can be uncovered in Melanesian practices, including 'practical' skills in house construction – estimating space, proportion, volume, load-bearing capacities, not to forget the strength of knots or the tension of ropes and building vines (Damon 2017). The Center has documented the considerable variety of PNG counting systems, and their operative patterns or frame pattern characteristics that link them to English language arithmetic. This is not the place to engage in the controversy over enumeration and arithmetic, on which Saxe (2012) has commented. (His own study over time of how Oksapmin counting practices have gradually changed with people's dealings in a modern currency of coins and notes, among other things, is exemplary.)

demonstration had been intended to show how digit is enticed to follow digit, one pulling another after it.

Hagen people accord a unique name to numbers 1, 2, 4, 8, other designations being combinations. In contrast with Iqwaye, who sequentially open up the hand starting with the thumb, Hageners start with the open hand empty, nothing yet grasped. Left hand first, fingers are closed down, one by one or two by two, into the palm. While I give Strauss's ([1962] 1990, 11, 228–229) original notation, the following description of the first five (one by one) includes my interpretation.

1. *Tenda (tsi)*. Little finger down. Being gathered into the base of the hand by being separated from the next finger, as half of a pair it leads into a further joining.
2. *Ragl*. Next finger down: the pair re-forms as 'one' pair [my gloss].[25]
3. *Ragltika* ('two-one'). The one pair now draws down another one, its descriptive implying co-presence.
4. *Tembokak*. The above one (single digit) was the result of a further separation and its other half – the index finger – follows. This concludes the initial sequence, resulting in 'one hand', *ki tenda* (4).
5. *Pömb tsi (tenda) gudl*. The 'one thumb' may be closed down on the fist as an extra, but upright and thus by itself it is 'out of place' or out of line. Closing it down does not make the 'hand'; completing 4 (two lots of two) fingers does this.

Hand seeks hand. The initial hand is only a half: the left must be complemented by the right in order to complete a whole 'one hand', *ki tenda (tsi)*, or *engak* (8) to give it the fourth number name. When counting items ceremonially, men proceed in twos; donors reckoning a line of shell valuables they are about to give to their exchange partners, for example, call out, '2, 2, [that's] 4; 2, 2, 4; 4, 4, 8'.[26] The impetus to complete or complement a one, and thus draw down a singleton's pair or 'partner', creates movement – an openness towards a new state of affairs, and then towards a further new state of affairs. I return in a moment to

[25]Grammatically, a pair takes a dual number or word form. 'One' here is my English language intervention, and glosses the Hagen pair as a 'completed whole', after Strauss ([1962] 1990, 11): 'Odd numbers are felt to be in need of complementation. Something by itself cannot really be counted, it is not complete, but like a half … [It is] "out of its place" … [They see] anything that stands alone as having "another half" somewhere, which it needs to complement it and make it whole'. A pair is verbally indicated either by naming two items together and adding the qualifier 'two' or specifying that each has its 'partner' or 'helper/supporter' (the same term used for exchange partners).

[26]And see A. Strathern (1977, 18) as well as Lancy and Strathern (1981, 783–784). Vilaça (pers. comm., 2019) has pointed out that in the procreation of persons it doubtless matters whether a pair is two of a kind (same sex) or two others (cross sex), only the latter being generative. We have seen that Iqwaye counting proceeds between ones and twos as alternately male–female entities. In Hagen, on the face of it, almost anything can be paired: pork and vegetables/birds and marsupials/dogs and marsupials/brother and brother/elder and younger [brother], while sets – making general categories, pairing things – maybe referred to as (joining) 'grandfather and grandson' or as 'husband and wife' (Lancy and Strathern 1981, 781, 789). Merlan and Rumsey (1991, 113–116, 241–242) offer a germane discussion of Hagen and Ku Waru pairing.

the effect of a terminus in the initial conclusion of the whole fist (itself half a hand of 8), such that one hand (4) is also a stopping point in bringing a double pair of digits together. Meanwhile, note that each digit, in turn, may also stand for a whole 8, and donors routinely count the number of eights they have assembled. To enhance a sense of magnitude, '5' (mobilizing the thumb) operates as a superior 4, '10' as a superior 8 with 'two thumbs down'. The thumb is also used numerically, on non-ceremonial occasions, for numbers beyond four. Thus 6 may be reckoned as '[a hand with] two thumbs out of place'; 7 '[a hand with] three thumbs out of place'.

'One' slips between different positions, then. In the movement from the little finger (one) as half a pair (one) to a fist with two pairs or four fingers down (one), to the free-floating thumb (one), they are all on the Hagen hand. The first fist closure is momentary, to be joined by the other fist. It is the ability *both* to get from one position to the next *and* treat a position as an end-point of previous positions that seems to facilitate sequential enumeration. – The fourth one is a 'leaky' oddity when needed; the thumb spilling over from the basic counting can indicate less or more than a hand of pairs. – Now the complete(d) *ki tenda* (8), Strauss ([1962] 1990, 228) observed, is itself said to look 'like a piece that has been cut off'. Recall that the end of one thing is the beginning of another, which is what a 'cut' accomplishes. If ceremonial presta-tions are prime occasions for counting, on the part of Hagen men at least, however many hands donors amass, they are only to be completed by the number of hands that will, in the future, be drawn as return gifts from their partners. That complementary return is, in effect, what will happen next.[27] Maenge (M. Panoff 1970, 360) deploy the same idiom of a 'cut' for a sense of a numerical stopping point (in their case a full 20, not 8), when 'no toe is left' to count. In short, I infer, the terminus marks the moment when the next action starts to take place.

In these various sketches from Melanesia, generations of plants and persons replace one another. The evidence (yams, children, valuables) may be joyously – or anxiously – enumerated. It is almost as though numbers were, themselves, a form of life, their liveliness evinced in the ever-forward pull of counting one thing after another, as well as a form of death, in the stopping places that over and again open up new sequences, allow fresh starts.

Acknowledgements

Following the stimulus of the original symposium, without which this would not have been written, I am indebted to Knut Christian Myhre and Anthony J Pickles both for their close reading of the text and for their critical interventions. The reverberations they set up go

---

[27]Pickles here adds that pairs are completed inside encompassing pairs, like the Maenge soul; in both cases the work of the social places pairs within larger pairs.

beyond the specific references made here. Dialogue with Aparecida Vilaça has been much to my benefit, and many thanks for the preview of her revised paper. A version was presented at the 2019 Academia Europaea annual congress in Barcelona, under the title 'Counting people'. This also comes with acknowledgement to the International Balzan Foundation for funding the research project, *Time and the ethnographic horizon in moments of crisis* (Centre for Pacific Studies, University of St Andrews), to which the substance of the article relates.

Disclosure statement

No potential conflict of interest was reported by the author(s).

References

Biersack, Aletta. 1982. "The Logic of Misplaced Concreteness: Paiela Body Counting and the Nature of the Primitive Mind." *American Anthropologist (NS)* 84 (4): 811–829.

Carrier, Achsah. 1981. "Counting and Calculation on Ponam Island." *Journal of the Polynesian Society* 90 (4): 465–479.

Coupaye, Ludovic. 2013. *Growing Artifacts, Displaying Relationships: Yams, Art and Technology Amongst the Nyamikum Abelam of Papua New Guinea*. Oxford: Berghahn.

Crook, Tony. 2007. *Exchanging Skin: Anthropological Knowledge, Secrecy and Bolivip, Papua New Guinea*. Oxford: OUP for The British Academy.

Damon, F. 2017. *Trees, Knots and Outriggers: Environmental Knowledge in the Northeast Kula Ring*. Oxford: Berghahn Books.

Goodale, Jane C. 1995. *To Sing with Pigs is Human: The Concept of Person in Papua New Guinea*. Seattle: University of Washington Press.

Ingold, Tim. 2019. "Art and Anthropology for a Sustainable World." *JRAI (ns)* 25 (4): 659–675.

Lancy, David, and Andrew Strathern. 1981. "'Making Twos': Pairing as an Alternative to the Taxonomic Mode of Representation." *American Anthropologist* 83: 773–795.

Lipset, David, and Jolene Stritecky. 1994. "The Problem of Mute Metaphor: Gender and Kinship in Seaboard Melanesia." *Ethnology* 33 (1): 1–20.

Matang, Rex, and Kay Owens. 2004. "Rich Transitions from Indigenous Counting Systems to English Arithmetic Strategies: Implications for Mathematics Education in Papua New Guinea," *Ethnomathematics and Mathematics education*, Proceedings of 10th International Congress of Mathematics Education, Copenhagen, edited by Franco Favilli, Pisa: Tipografia Editrice Pisana.

Merlan, Francesca, and Alan Rumsey. 1991. *Ku Waru: Language and Segmentary Politics in the Western Nebilyer Valley, Papua New Guinea*. Cambridge: Cambridge University Press.

Mimica, Jadran. 1988. *Intimations of Infinity. The Mythopoeia of the Iqwaye Counting System and Number*. Oxford: Berg.

Moutu, Andrew. 2013. *Names are Thicker Than Blood: Kinship and Ownership Amongst the Iatmul*. Oxford: OUP for The British Academy.

Myhre, Knut Christian. 2016. "Membering and Dismembering; The Poetry and Relationality of Animals Bodies in Kilimanjaro." In *Cutting and Connecting: 'Afrinesian' Perspectives on Networks, Relationality, and Exchange*, edited by K. C. Myhre, 114–131. Oxford: Berghahn.

Panoff, Françoise. 1969. "Some Facets of Maenge Horticulture." *Oceania; A Journal Devoted to the Study of the Native Peoples of Australia, New Guinea, and the Islands of the Pacific* 40 (1): 20–31.

Panoff, Françoise. (1970) 2018. *Maenge Gardens: A Study of Maenge Relationship to Domesticates*, edited by Françoise Barbira-Freedman. Marseille: pacific-credo Publications.

Panoff, Françoise. 1970. "Food and Faeces: A Melanesian Rite." *Man (n.s)* 5 (2): 237–252.

Panoff, Michel. 1968. "The Notion of Double Self among the Maenge." *Journal of the Polynesian Society* 77 (3): 275–295.

Panoff, Michel. 1970. "Father Arithmetic: Numeration and Counting in New Britain." *Ethnology* 9 (4): 358–365.

Pickles, Anthony. 2019. *Money Games: Gambling in a Papua New Guinea Town*. Oxford: Berghahn.

Pospisil, Leopold. (1963) 1978. *The Kapauku Papuans of West New Guinea*. 2nd ed. New York: Holt, Rinehart and Winston.

Rumsey, Alan. 2000. "Agency, Personhood and the 'I' of Discourse in the Pacific and Beyond." *JRAI* 6 (1): 101–115.

Saxe, Geoffrey B. 2012. *Cultural Development of Mathematical Ideas: Papua New Guinea Studies*. Cambridge: Cambridge University Press.

Scaglion, Richard. 1999. "Yam Cycles and Timeless Time in Melanesia." *Ethnology* 38 (3): 211–225.

Strathern, Andrew. 1972. *One Father, One Blood*. Canberra: ANU Press.

Strathern, Andrew. 1977. "Mathematics in the Moka." *Papua New Guinea Journal of Education* 13 (1): 16–20.

Strathern, Marilyn. 1972. *Women in Between: Female Roles in a Male World*. London: Seminar (Academic) Press.

Strathern, Marilyn. 2019. "A Clash of Ontologies? Time, Law and Science in Papua New Guinea, HAU." *Journal of Ethnographic Theory* 9 (1): 58–74.

Strauss, Hermann (with Herbert Tischner) (1962) 1990. *The Mi-culture of the Mount Hagen People, Papua New Guinea*, edited by G. Stürzenhofecker and A. Strathern, translated by B. Shields. Department of Anthropology, University of Pittsburgh: Ethnology Monographs.

Telban, Borut. 1998. *Dancing Though Time: A Sepik Cosmology*. Oxford: Clarendon Press.

Vilaça, Aparecida. 2019. "Inventing Nature: Christianity and Science in Indigenous Amazonia." *HAU: Journal of Ethnographic Theory* 9 (1): 44–57.

A pagan arithmetic: unstable sets in indigenous Amazonia*

Aparecida Vilaça

ABSTRACT

In the Amazonian literature, the scarcity of numerical terms and lack of interest in counting, which characterize diverse indigenous peoples in the region, are usually associated with linguistic issues or cultural limitations. My purpose in this paper is to take a different approach and relate the enumeration processes of one of these peoples, the Wari', to their conception of beings as intrinsically transformative, which affects the determination of quantities and the stability of sets.

In the field of numerology, as in other fields, the spirit of each system must be determined independently of the observers' own categories. (Lévi-Strauss 1978 [1968]: 335)

The stones resisted arithmetic as they did the calculation of probability. Forty discs, divided, might become nine; those nine in turn divided might yield 300. (Borges, 'Blue Tigers')

Mathematics in the confrontation zone

Much has been said about the 'culture shock' resulting from interethnic contact, more specifically the contact between native and European peoples, which, among other things, generates humiliating situations for indigenous populations (Robbins 2004). Approaches to these confrontations tend to focus primarily on the religious sphere, bodily and alimentary practices, and kinship (especially marriage rules). It seems to me, though, that a central aspect of the conflicting views tends to pass unnoticed: namely, mathematics.

---

*I am grateful to the many people who contributed to the discussions that led me to the final version of this article, starting with the participants in the seminar Science in the Forest, Science in the Past II, which took place at the Needham Institute of the University of Cambridge (UK) in May 2019. Among them, I especially thank Geoffrey Lloyd and Willard McCarty, who organized it together with me, and who, with Stephen Hugh-Jones, were the first readers of the text. I owe Marilyn Strathern important ideas for the development of this work, as will be clear to the reader. João Biehl and Fernando Codá Marques made essential comments when I presented the work at the Brazil Lab at Princeton University in February 2020, and Anne-Christine Taylor and Evelyn Fox-Keller reviewed their penultimate version, offering ideas for its rewriting.

In the case of Amazonian peoples, this problem is especially relevant since they have become famous in both the anthropological and linguistic literatures for their poverty in numbers and, consequently, for their disinterest – translated by lay people and scholars alike as their difficulty – in counting. While indigenous peoples obviously did not see the peculiarities of their (a)numeric system from the perspective of lack, the situation changed with the establishment of stable contact with white people, who began to impute to them a self-image of generalized poverty, including the strange idea of poverty in mathematical reasoning.

On the side of the whites, the response to this observation was varied: traders exploited the situation to profit from financial transactions, while teachers, many of them missionaries, took pity and decided to teach them numbers and counting, along with the word of God.

My objective in this paper is to bring mathematics, more specifically, arithmetic, to the centre of the discussion on the contact between indigenous and white people in order to emphasize the peculiarities of the former's mathematical thought, associated, I believe, with their conceptions of what relations do to sets.

Moral calculations and the lack of numbers

A first characteristic that scholars of indigenous Amazonian mathematics emphasize is the inclusion of moral and relational questions in their calculations, a factor that becomes especially evident in the school context.

The works of ethnomathematics, particularly those of Mariana Ferreira among Amerindian peoples, show how such questions are often implicated in the ideally precise calculations of mathematics. Among the Xavante of Central Brazil, for example, where she worked as a teacher, the anthropologist witnessed the elaboration of mathematical problems fairly peculiar by the children, with answers equally unusual in the universe of arithmetic precision. The problems and their answers, given by two children aged 8 and 9, respectively, were as follows (Ferreira 1998, 85–86 see too D'Ambrosio 1990, 1994):

1) *My father is going to hunt paca. He has a box of cartridges. How many pacas will he kill? Answer: He will kill 3 or 7 pacas, however many he manages to kill.*

2) *In my father's swidden there is a lot of maize. My mother is going to make maize cake. How many cakes will she make? Answer: She will make 3 very large cakes, for everyone to eat.*

According to Ferreira (1998), in both problems, 'it is clear that there is no strict relationship between the quantity of maize cobs or rifle cartridges and the quantity of cakes made or pacas killed, respectively. The solutions to these problems involve other relations that are not included in the mathematical problems.'

It should be noted, as Lave (1988) demonstrated in her research on the calculations made by US adults when shopping in supermarkets, that the influence of non-mathematical principles on the formulation of problems and their results is not a characteristic exclusive to indigenous peoples. Unlike the supermarket-shopping Americans, however, the moral calculations made by Amerindian peoples elicit a general negative view of their cognitive capacities. The 'erroneous' calculations are, in the view of lay white people (and some academics, as we shall see later), a natural consequence of the lack of numbers and counting systems. Diverse native Amerindian languages are known to have no specific terms for numbers, while those that do are often limited to just a few numbers (very often possessing terms for 1 and 2 only), which would suggest the inability of these populations to quantify and add up. Although some Amerindian peoples are known to exhibit complex base 5, 10 and 20 systems, including the Palikur studied by the missionary linguist Diana Green (1994), here I shall concentrate on systems that became known in the literature as 'rudimentary' (see Lévi-Strauss 1978, 336), since these include that of the Wari', a people living in southwest Amazonia, whom I have been studying for some decades and who will provide the bulk of the ethnographic data analysed here.

According to the classification made by Green (1997) in a survey of the numerical systems of 45 indigenous languages in Brazil, the Wari' quantification system, though not included in her survey, would be base 1, just like those of the Pirahã, Canela, Ashaninka, Kulina, Tenharim, Sanumá and Nadeb (Maku) (Green 1997, 4). Along with base 2 systems, these systems are, the linguist argues, associated with a 'global or holistic thought,' whose quantifiers are associated with the 'total context or the notion of totality' (Green 1997, 7), contrasting with the 'analytic and synthetic' thought associated with base 10 and 20 systems (Green 1997, 8). This classification is illustrated as follows: 'A man does not say: "I'm going to chop down eight posts to make the house." He says: "I'm going to chop down a post for each corner and one more for each side." And if someone asks him how many he is going to chop down, he will reply: "I'm going to chop down several" (Green 1997, 4). Describing these systems, Green (1997, 7) continues:

> Even the meaning of the few terms utilized is not well-defined; it is very common for the term for 2 to mean 'some' and the term for 3 to mean 'many,' since they are relative to the total [...] The Canela language, for example, has no specific numerical terms; it is limited to general terms such as 'only,' 'a pair,' 'some' and 'many.'
> (Green, 1997, 3)

Just like the Canela cited by Green, the Wari' do not have specific numerical terms, and the quantifiers are limited to the unit, *xika' pe,* which signifies 'alone,' and to the pair, *tuku karakan,* 'one facing the other' or *tokwan,* one of the terms for 'many.'[1] Above 2, they use terms for few or many, always relative, since they depend on the relational context, which is variable. For example, if hunters in a small village kill two peccaries, they will refer to their prey as 'many,' but if they live in a densely populated village and have to distribute the game to many, they killed a 'few' (see Vilaça 2019). It seems to me that this mathematical proportionality, which for Green is related to a notion of totality, is less important than the intentionality involved in the calculations: qualifying a number as many or few depends on what one wishes to emphasize or render invisible. Every anthropologist who has lived with an indigenous people knows perfectly well that the qualification of the product of a gathering or hunting expedition as 'many' or 'few' is related to the interest in sharing more or less. The calculation of kin also involves contextual sets: my older Wari' friends, who never went to school, used to give different answers each time I, recently arrived for a visit, asked them how many grandchildren they had. Whether they included grandchildren who lived faraway in the (nominal) list or omitted them depended on whether the latter visited frequently.[2]

Although they do not have nominal quantifiers beyond 2, which for its part may signify 'many,' the Wari' can indicate larger quantities, up to 10, with their fingers, even without naming them, as Green (1997) and Gordon (2004, 497) also observed for other systems of this type. However, as happens among the Pirahã studied by Gordon (2004) and Everett (2005), who also do not quantify nominally beyond 1 and 2, the Wari' demonstrate no interest in expressing totalities when making an equivalence between the objects or persons cited in their narrative and their fingers. Even if sometimes they unite all the selected fingers in order to show a total amount, they soon lose interest in the sequence, often repeating the same fingers, or frequently using just one hand (see the same for the Munduruku in Pica et al. 2004, 503) As Evelyn Fox-Keller has suggested (personal communication, 2019), finger touching is above all an act of particularization rather than counting.[3]

---

[1] Daniel Everett and Barbara Kern translate the 'quantifiers' of the Wari' language as follows: *paric* ('to be little'), *pije* ('child, to be little'), *tocwan* ('to be many'), *tamana* ('to be numerous, many'), *iri'mijo* ('to be many'), *xam* ('to be complete, everything') and *pi'pin* ('to finish completely') (1997, 349).

[2] According to Silva (2006, 78) in a study of Guarani and Kaiowá mathematics, 'in the indigenous view, numerical precision is not a cultural value.' Simon Schaffer (personal communication, 2019) suggests that we should differentiate precision from accuracy. In that case, we might say they are precise but not accurate. Obviously, we cannot generalize this observation to indigenous Amazonia as a whole. As S. Hugh-Jones has observed (personal communication, July 2018) for the Barasana (Tukano, Upper Rio Negro), counting is a cultural value, especially in the context of numbered shamanic songs, architecture and astronomy. The same occurs among the indigenous peoples of the Upper Xingu (Fausto, personal information, 2018).

[3] In a collective work from the intercultural university of Rondônia, *Teias do conhecimento intercultural* (Leite 2013), with chapters composed by diverse indigenous students and containing texts in Portuguese and in the indigenous languages, the author of the article titled 'Forms of counting' in Portuguese translates the fact that the Wari' knew how to count as follows: *'taxi nanain ka xat kaka* = they knew where to settle/stop *(xat)'* (Oroat 2013, 170–171). In another article in the same work, the title in Portuguese, 'Traditional form of counting of the OroNao' people,' is translated into Wari' as: 'How the OroNao' say how many things there are in their true language.' Interestingly, the word for 'how many,' *kain,* also means 'how,' which seems to be its primary sense. Some drawings of arrows represent the numbers 1, 2, 3 and 4. The caption to the drawing in Portuguese, 'Numerical terms

With the introduction of the school among the Wari' at the beginning of the 1970s, founded by the same US Evangelical missionaries who catechized them, they learnt the numbers in Portuguese for school work and for commercial transactions with white people. However, following their access to the so-called intercultural teaching at the start of 2000s, relativist in approach, teachers – especially those in secondary and university education – began to ask them to translate numbers into their own language, a sign of their respect for the different indigenous cultures present in the classrooms. Without being able to assess the effect of this on the other indigenous peoples present in the multi-ethnic classrooms, I observed considerable embarrassment on the part of the Wari,' who suddenly saw themselves as poor in numbers. They began, then, to try to settle on names in their language to signify other numbers beyond 1 and 2. The term for few, *parik,* became the number 3. Thereafter, they started to use, somewhat randomly, different terms for 'many' and 'full' to designate all the numbers from 4 to 10. This type of numbering remained confined to the classrooms and even there, with the exception of 1, 'alone,' numbers are expressed in Portuguese when they do sums.

The invention of names for numbers is not a phenomenon limited to the Wari'. A recent article, co-written by a Pàrkatêjê (Gavião, Pará) student and his non-indigenous teacher, describes this process of invention, which the authors call an 'amplification and creation of the written and phonetic forms of numbers in Pàrkatêjê,' based on public discussions with the community, including the approval of elders and leaders (Valdenilson and Lima 2017, ms; also see Tenório, Ramos, and Cabalzar 2004, on the invention, by the Tuyuka, of 'large' numbers in their own language).

Among the peoples 'poor in numbers,' the Amazonian case of the Pirahã (Mura language, Amazonas) became famous in the literature, provoking an intense debate among anthropologists and linguists. Although studies based on cognitive tests (Gordon 2004) and linguistic analyses (Everett 2005) are controversial and have been subject to a variety of critiques (see Levinson 2005, 637–638; Wierzbicka 2005, 641), it is worth commenting on them briefly, given the similarity to the ethnographic data on Wari' arithmetic. In the case of the Pirahã, not only are the existing terms limited to 1 and 2, going from there to 'many,' but even the term for 1 is not 'stable,' since it is also used to signify small, acquiring the meaning of 'roughly 1,' a category nonexistent, as

---

of the OroNao' people,' is translated as 'How the OroNao' say how many *karawa* there are [*karawa* = animal/prey, non-human]' (Mauricio Oronao' 2013, 119–120). In Hymn 88 of the bilingual hymnbook, however, it is stated in Wari' that only God knows how to count the stars, using the Portuguese verb *contar*. Finally, in another hymn, 113, the term 'count' in the Portuguese phrase 'They were 100 sheep, but one afternoon, on counting them all ... ' is translated into Wari' as to 'look/search' (*noro*): 'He searched and searched and just one did not exist.' Zero appears in just one of the texts, produced by a university student, who translates the term as 'to not exist' (*om na*) (Ororam Xijein n.d., 26). Among the Yanomami, according to the dictionary produced by Lizot (2004, 290) the term for to count, to evaluate a quantity, to count with one's fingers – *owë!* – is the same as to imitate or equal.

Gordon (2004, 498) observes, in our system of whole numbers (also see Frank et al. 2008, 820).[4]

Notably, Gordon and Everett venture distinct hypotheses for the absence of numbers among the Pirahã: for the former (Gordon 2004, 496; 499), it is related to the absence of nouns designating precise quantities; for Everett (2005, 634; 643), though, the question is not one of linguistic determinism. Rather he attributes the absence of numbers to 'cultural constraints,' more precisely to the 'inability in principle to talk about things removed from personal experience' (Everett 2005, 633).[5] Unlike the Wari', however, the Pirahã, according to Everett, were unable, even after eight months of continuous lessons from the missionary linguist from the Summer Institute of Linguistics (SIL) and his wife, to learn the numbers in Portuguese or even to complete the most basic sums, such as $3 + 1 = 4$ (Everett 2005, 625–626).[6]

The Munduruku case (Pica et al. 2004), which became equally well known, is also worth mentioning, although – differently to the cases discussed here – they exhibit a base 5 system in which the numbers 3–5 are composed from the terms for 1 and 2 (see too Green 1997, 8, 9). According to Pica et al. (2004, 500), however, although they can nominate up to 5, they do not go beyond 2 to specify quantities, not differentiating $n+1$ after 3 or 4, and may use the term for 5 for quantities of 5, 6, 7, 8 or 9 elements, for example:

> Above 5, there was little consistency in language use, with no word or expression representing more than 30% of productions to a given target number. Participants relied on approximate quantifiers such as 'some' (adesu), 'many' (ade), or 'a small quantity' (burumaku). (Everett 2005, 500)

The authors' conclusion is particular interesting here: 'This 'crystallization' of discrete numbers out of an initially approximate continuum of numerical magnitudes does not seem to occur in the Munduruku' (Everett 2005, 503). I shall come back to this idea of a lack of 'crystallization,' as well as the qualification, by both Gordon and Everett, of the 1 of the Pirahã as 'roughly 1.'

---

[4] In his words: 'Of particular interest is the fact that the Pirahã have no privileged name for the singular quantity. Instead, "hói" meant "roughly one" or "small," which precludes any precise translation of exact numerical terms' (Gordon 2004, 499). See D. Everett (2005, 623) on the imprecision of the unit.

[5] In relation to cultural questions, it is worth noting that Green (1997), in her comparative study, rejects any association between numerical restriction and material poverty. The author notes that some people with a numerical restricted terminology live in a material universe as rich as that of people with more complex systems and could, when interested, adopt a more exact form of counting for particular objects like those coming from the white world, such as cartridges and batteries (Green 1997, 7). In relation to cognitive questions, a large comparative mathematical project, conducted among 10 different indigenous peoples of Papua New Guinea, concluded that: 'No direct link between counting systems and cognition was found but connections to indigenous classification systems were' (Lancy 1981, 450); 'basic numeracy appears to be totally unrelated to cognitive development' (Lancy 1981, 447).

[6] In the follow-up to this debate, years later, Caleb Everett (Everett, C. and Madora ms), Daniel Everett's son, performed new tests among the Pirahã with the aim of refuting the data of Frank et al. (2008) and supporting those of Gordon (2004) due to the fact of having no success in observing 'one-to-one matchs,' reaffirming their supposed inability to recognize quantities higher than three (Gordon 2004, 15, 16).

The instability of 1 and the 'blue stones' of Borges

As I observed earlier, the Wari' term for 1, *xika pe'*, has the meaning of alone, which in Wari' sociocosmology points to a lack, not to the unit in itself. Someone is alone because they lack an other, a spouse, a relative. A single killed game animal is a sign of the hunter's lack of ability or luck: he lacks other preys. The most precise mathematical translation for alone would be – *n*, therefore, not the unit, such that the value of 1 is, as among the Pirahã, 'approximate,' always pointing to an absent 'other.'

The negative value given to 1 is not a characteristic exclusive to the Wari'. It is common to various lowlands peoples, as diverse as the Carib-speaking Ingarikó of Roraima (Amaral, personal communication 2017), the Tupi-Guarani-speaking Parakanã of Pará (Fausto, personal communication 2017) and the Barasana (Tukano, Upper Rio Negro), for whom the 'unit is incomplete and dangerous' (S. Hugh-Jones, personal information 2016). The relation between the unit and solitude is also evinced among the Krenak (Medeiros and Miranda 2009, 3948), the Canela (Green 1997) and among the 'Chiquitas' (probably the Chiquitanos of the Bolivian Chaco) according to the pioneer work of Conant 2013, 583; cited in Nykl 1926, 584), who observes that no more than a few tribes encountered in the region can count higher than 1 and 2 (Nykl 1926).[7] Similarly, Ferreira, in a study of the numerical concepts of the Xavante, relates the unit to solitude, contrasting the positive value given to 2 and to even numbers in general (Ferreira 2001, 91). Among the Yanomami, who also only have terms for the unit and the pair, passing afterwards to 'many,' 1 can be expressed both by the term for 'alone,' *yãmi*, which has the synonym of 'few' (Moura, personal communication 2019), and by the term for 'almost,' *mori*, in an interesting approximation to the notion of 'roughly 1' of the Pirahã (see Lizot 2004, 238; Moura and Kelly, personal communication).[8]

Another case of interest is that of the Araweté, whose term for the unit, *tisipe*, is also the term for alone, a quality also expressed by *jije*, 'without anything,' 'without,' 'alone' (Caux 2015, 35). *Jije* indicates a lack, the dissociation of a connected element: to eat meat without a plant-based complement and vice-versa is to eat *jije*, since they should always be eaten together (36). It can also signify 'whole,' but only in relation to something that could be split and divided. A game animal brought back whole to the village is *jije* since it will be distributed

---

[7]Medeiros and Miranda (2009, 3948) argue apropos the Krenak: 'Many languages use words like "only, alone" to express unitary quantity, as is the case of Krenak itself, in which the word *putfik* was recorded by all researchers as corresponding to the noun for the numeral quantifier "one" and to the meaning "only, alone." This is also the case of various other Brazilian indigenous languages (Aryon Rodrigues, personal communication).' For a non-Amazonian context, Urton and Llanos (1997, 78), in his study of the ontology of Quechua numbers (Sucre, Bolivia), writes that 'the motivation of the two is the solitude (*ch'ulla*) of the one.'

[8]The solitary character of 1 is equally emphasized by Ubiratan D'Ambrosio (2015) in the preface to the book by Mariana Ferreira (2015). See Lancy and Strathern (1981, 783–784) on the place of the pair among the Melpa of Papua New Guinea.

(38), but a small piece of game is 'non-*jije*,' since it cannot be divided further (40), just like a packet of sugar. The author concludes:

> These are, it seems, two different meanings of the same word. What is interesting is that, though in opposite directions, both appear to be founded on the same relation: one that informs an inadequate dissociation (like that of the hunter without his game), another that speaks of a separation still to be realized (like that of the whole game). One is described by the interval between two terms that should be united; elements that should be together but are separate. Another by the continuity between parts that should be separate; elements that are whole but should be distributed. (Caux 2015, 43)

The Araweté data thus insert a temporal dimension into the unit: it is provisional, destined to be either complemented or divided. Allow me, then, to make an approximation with Strathern's analysis (this volume) on the meaning of 1 for the Hagen. According to the author, 1 exists both as a part dissociated from 2, and in this sense indicates a lack, and as the sum of different units that form a whole of another kind (a hand, for example), which, just like with the Araweté, can be divided once again, thus containing a multiplicity. This implies that, as in Amazonia too, 1 is essentially unstable, just like 2, destined to become 1 (1 pair, 1 couple) through the insertion of a third unitary term (a child, for instance). It is the constant shift from one term to the other that characterizes these systems; the focus is on the transition, not the fixity. In this sense, defining a certain quantity as discrete or not depends on its momentary state in the process of transformation. In a previous article (Strathern 1992), the author analyses the production of the unit through the context of the exchanges: rather than existing *a priori*, it is produced at the moment when they are elaborated, based on a game of persuasion and influence that produces an equivalence between the items set to be exchanged. Criticizing the anthropological analysis of barter, which takes the identity of the objects beforehand, the author writes ' … things are thereby presumed to present a unitary quality […] The identity of the units already exist "in nature" […] The discreteness of things […] seem evident enough. Yet their abstraction as units belongs to a particular cultural practice which assumes the priority of individual identity' (Strathern 1992, 172). She adds: ' … the unit appears as a metaphor for substitutability, it indicates an analogy between two items' (Strathern 1992, 176).

As I observed earlier, it is interesting to note that although the schooled Wari' have adopted the names for numbers in Portuguese, they remained monolingual for 1, always using the term for 'alone,' a fact also observed by the missionary linguist Barbara Kern (Everett and Kern 1997, 347). In a previous work (Vilaça 2018), this peculiarity of the numerical system currently used by the Wari' prompted me to suggest the incompatibility between their thought and the notion of unity objectified in the number 1, so central to Christianity.

The idea of '1' is a core notion for the fundamentalist Evangelical missionaries who catechized the Wari'. In the Bible, which they take literally, they identify various passages that extol the value of 1 and unity, either directly or indirectly through a critique of duality, as appears in the following verse from the books of Proverbs (20: 10), virtually an explicit course in moral mathematics: 'Double measures and double standards are an abomination to the Lord.'[9]

The opposition is not original and indeed I was led to it via Lévy-Bruhl (1985, 181–223) and his chapter on numeration in *How natives think*.[10] In his words:

> The 'unit' has maintained a prestige upon which the monotheistic religions and the monistic philosophies plume themselves. 'Duality' is often the antithesis of unity by qualities which are diametrically opposite … Many languages still preserve in their vocabulary traces of this opposition; and we speak of 'a double life,' 'duplicity,' etc. (1985, 209)[11]

The opposition between the unit and the pair was also noted by Pierre Clastres (1989) in a chapter of his most famous work *Society Against the State*, 'Of the One Without the Many,' based on the Guarani ethnography. Clastres attributes the valorization of the two or double in opposition to the one to the pregnancy of an anti-identity ethos among indigenous peoples. In his words:

> The imperfect earth where things in their totality are one is the reign of the incomplete and the space of the finite; it is the field of strict application of the identity principle. For, to say that A = A, this is this, and a man is a man, is to simultaneously state that A is not-A … To name the oneness in things, to name things according to their oneness, is tantamount to assigning them limits of finitude, incompleteness. (Clastres 1989, 173; see also Lima 1996, 1999, 2008)

Incompleteness is not due, therefore, as we tend to think, to the openness of the set, but in its delimitation and closure.

The one and the double are central themes in the work of Lévi-Strauss (1995), particularly explicit in his analysis of a set of myths related to twins. The myths drew his attention to the difference between twins in Indo-European mythology – who, though initially different, are eventually conceived as identical – and twins in the myths of the Americas, who become increasingly different over the course of life. The author concludes therefore that, in relation to the Americas, 'all unity contains within itself a duality' (Lévi-Strauss 1995, 64).

---

[9]According to Alexandre Koyré, God created the world 'on number, weight and measure' (1980, 70). On mathematics as an ideal instrument for the description of nature, see Tabak (2011).

[10]See Ascher and Ascher (1969), Carrier (1981), Biersack (1982), Mimica (1988), Ascher (1991, 2002), Urton (1997) and Verran (2001); see Lave (1988) for a western context.

[11]According to Urton and Llanos, the Quechua consider that 'the condition of "one/odd" represents a negative, problematic and unhappy state of things, while the "two/pair" represents a positive, happy and eventually productive state of things' (1997, 57; also see 58–61). For an exception to the negativity of the one, see Mimica (1988, 46) on the valorization of the unit among the Iqwaye, who see totality as indivisibility and thus positive (Mimica 1988, 47).

This is not, however, a simple duality but a 'dualism in a perpetual state of dis-equilibrium' (Lévi-Strauss 1995, 235), whose dynamic had been prefigured almost 40 years earlier in the author's analysis of the relation between diametric and concentric dualist systems in 'Do dual organizations exist?' (Lévi-Strauss 1963 [1958], 132–163).[12]

I cannot refrain from observing, without any mathematical basis, but simply for sheer aesthetic pleasure, as the French master would say, that this image of dualism in disequilibrium described by himself is reminiscent of some of the numerical systems described here. They valorize the 2 or the pair, but do not pass from 2 to 3 and 4, as occurs in binary counting systems, sustaining the 2 indefinitely, albeit in ever distinct configurations, which maintain a fractal relationship between themselves.[13] What matters, as Lévi-Strauss reminds us in 'The Story of Asdiwal' (Lévi-Strauss 1983), is that although the distance between the terms tends to become smaller and smaller, the paired elements never become identical, lest they cause the paralysation of the world (Lévi-Strauss 1995 [1991], 208–209), which takes us back to the danger of the 1 evoked by the Barasana and echoed by Clastres's view cited earlier.[14] In my view, this incessant reproduction of pairs, which move closer without ever becoming equivalent, provides us with an interesting image of the infinite, which a Wari' young man defined in the classroom in the following form: 'never ceases to be far away.' Among them, while 2 does not become 4, neither does it turn into 1.[15]

Not just Christian morality properly speaking valorizes unity. As Vernant (1972, 95–102) observed, the Greek geometry and mathematics that gave rise to modern science of the kind taught in classrooms, developed through the fixing of a point of view for observing celestial bodies, whose orbit became mea-surable and predictable in a world conceived in spherical terms (see also Caveing 1990 and Seidengart 2006). By contrast, diverse Amazonian universes, including those of the so-called poor in numbers, the Pirahã, studied by Marco Antônio Gonçalves, are composed of superimposed layers and, therefore, largely incompatible with the unification of perspective implied in the notion of centre. In these universes, each different layer constitutes a world apart,

---

[12]See Lagrou (2007, 142–155) on the place of unstable dualism in Kaxinawá art, in the form of the figure-ground opposition. See Almeida (2015) and Hugh-Jones (2012) on the 'concrete' mathematics of Amerindians.

[13]An example of this fractal schema can be found in Lévi-Strauss (1995, 55), developed further in Viveiros de Castro (2000, 24–30, 2001, 28–33); also see Viveiros de Castro (2012).

[14]In 'The story of Asdiwal', Lévi-Strauss writes: 'The above two schemata are integrated in a third consisting of several binary oppositions, none of which the hero can resolve, although the distance separating the opposed terms gradually dwindles' (Lévi-Strauss 1967, 19). Also see the 'binary operators' in The Naked Man (Lévi-Strauss 1981). See Viveiros de Castro (2000, 24–26, 2001, 28–33) on the 'imparity of the two.'

[15]According to Silva (2006, 73), the Guarani and Kaiowá of Mato Grosso define very large quantities as those that 'continued further and further, or were denominated those without an end.' An image of the infinite, though not made explicit as such, also emerges in the description of the shaman Davi Kopenawa of the xapiri spirits as shining, miniscule and brilliant dust motes (Kopenawa and Albert 2013, 55–61). This infinitude is associated with the multiplicity of pairs when the shaman, seeking to describe them, compares them to his image in a pair of hotel mirrors: 'I was alone before them, but at the same time they showed a lot of identical images of me' (Kopenawa and Albert 2013, 61). See the analysis of these images in Viveiros de Castro (2007).

inhabited by beings with distinct viewpoints whose interaction involves a complex play of perspectives (Gonçalves 2001, 167–175; 407).[16]

While the association between mathematics and culture or ontology is undeniable, I tentatively propose a different connection to those made by Everett (2005) and Green (1997). The former, we may recall, makes the association through the idiom of lack: few numbers, imprecision and difficulty in counting are, he argues, related to the disinterest in everything that goes beyond immediate experience (Everett 2005, 623), which seems odd when we learn about the complex cosmology of the Pirahã and the importance of (invisible) spirits in everyday life (Gonçalves 2001). For Green (1997, 7), the association is positive: peoples with few or no numbers demonstrate holistic thought, focused on the totality.

While the objection to Everett's simplistic culturalism is clear in the criticisms of the commentators to his article (Everett 2005), Green's holism should be questioned precisely because of the incompatibility between the idea of totality and the relational thought that characterizes the Wari' and many other Amazonian peoples like the Pirahã and the Canela, to cite only those mentioned here. The quantities are determined by the context of the interaction between, on the one hand, the people and the objects counted, and, on the other, the perspective of the one who counts. This relational thinking evidently refers to the so-called Amerindian perspectivism, based at the coexistence of multiple worlds that never overlap or add up (see Viveiros De Castro 1996, 1998; Lima 1996, 1999; Vilaça 2002, 2005, 2009). Although it is not possible to make a direct connection between perspectivist thinking and the lack of interest in enumeration and counting, considering that other equally perspectivist peoples, like the Palikur and the Kaxinawa, for example, have more precise counting systems, it is worth considering that the idea of a lack of 'crystallization' (to use the expression of Pica et al. 2004) of larger numbers may be present among them, as it happens regarding numbers above five for the Munduruku. To advance with this hypothesis, though, we need detailed ethnographic descriptions of enumeration and counting systems of Amazonian peoples, which are, up to the present, very few. It seems reasonable, however, to consider that multiple and unstable worlds point to an interest in maintaining the sets open and quantities unstable, something that matches Viveiros de Castro's observation, in his inaugural article on perspectivism, concerning the epistemological choices of Amerindians: 'we are faced here with an epistemological ideal that, far from seeking to reduce "ambient intentionality" to its zero degree in order to attain an absolutely objective representation of the world, instead makes the opposite wager' (Viveiros De Castro 2013, 25). This echoes with Roy Wagner's observation (1975: 88)

---

[16]See Wagner (1975, 130) on the idea of centralization of viewpoints as part of the historical trajectory of the modern sciences.

on the action of non-western peoples constituting a 'continual adventure in 'unpredicting' the world.'

Rather than say that they cannot count, we could say that, in traditional contexts, they do not want to count, given that, as we saw in the answers to the school problems presented above, and as Green (1997, 7) shows, indigenous peoples rapidly manage to define precise quantities when dealing with objects of external origin and in transactions with whites, whether or not they adopt western numeration.[17] The crucial mathematical question, however, as I shall discuss below, is the indefinition of 1, which we saw is related to the negative value attributed to the unit, a fact that, in the constitution of a numerical system like ours, whose elements are composed by $n+1$, has important consequences.

An article co-authored by a mathematician and a philosopher (Nirenberg and Nirenberg 2018), suggestively titled 'Knowledge from pebbles. What can be counted and what cannot,' explores the question of the instability of the constitutive elements of sets in a very interesting way. The authors base their idea around a tale by Borges called 'Blue tigers,' which I shall provide a summary based on Borges's (1998) original text, somewhat extended given how much it enchanted me.[18]

A Scottish professor of logic, teaching at an Indian university, lives in a world founded on rationality, save for his obsession with tigers, which populate his dreams. One day, he learns of a report that in a village in the region of the Ganges were found blue tigers, which he would have ignored had it not been for the fact that the tigers in his dreams had turned blue. Consequently, he decides to go to the village to encounter the tigers, but the local inhabitants, though pretending to guide him, never put him on the right path. One night while everyone is asleep, he decides to explore alone a plateau that they had told him was prohibited, since the gods turned all those who venture there blind or mad. Suddenly he sees a bright blue colour in a crevice in the ground, exactly the same colour as the tigers in his dreams. They are pebbles, smoothly rounded and regular in size, like those used to count. He places a small handful in one of his pockets and returns home. When he puts his hand in his pocket to remove the pebbles, he notes that there were now more than the amount he had collected: 'I would look fixedly at any one of them, pick it up with my thumb and index finger, yet when I had done that, when that one disc was separated from the rest, it would have become many.' Henceforth the philosopher begins to closely observe these stones that multiply or diminish without any logic: 'I picked up the discs, raised them high, dropped them, scattered them, watched them grow and multiply or mysteriously dwindle.' And he continues: 'There are mathematicians who say that three plus one is a

---

[17] On the coexistence of two different counting systems among Melanesian peoples, see Carrier (1981, 466).
[18] I deeply thank Willard McCarty for suggesting (and sending) me the articles by those authors.

tautology for four, a different way of saying "four" ... But I, Alexander Craigie, of all men on earth, was fated to discover the only objects that contradict that essential law of the human mind [...] If three plus one can be two, or 14, then reason is madness [...] Naturally, the four mathematical operations – adding, subtracting, multiplying, and dividing – were impossible. The stones resisted arithmetic as they did the calculation of probability. Forty discs, divided, might become nine; those nine in turn divided might yield 300 [...] As I manipulated the stones that destroyed the science of mathematics, more than once I thought of those Greek stones that were the first ciphers and that had been passed down to so many languages as the word "calculus." Mathematics I told myself, had its origin, and now has its end, in stones. If Pythagoras had worked with these ... '

Nirenberg and Nirenberg use Borges's powerful image to critique the set theory that founds mathematics. According to the authors (Nirenberg and Nirenberg 2018, 3), if we ask a mathematician what counting is, he or she will reply that 'to count a finite set is to assign to its elements, in a one-to-one manner, the numbers 1, 2, 3, ..., n, without missing any one of the latter.' Everything is given previously, such that this type of problem begins with the phrase: 'given a set A,' or 'given the elements of the set A,' and 'given the natural numbers 1, 2, 3, 4, and so on' (Nirenberg and Nirenberg 2018). In the authors' words: 'We assumed that there is no question as to what elements belong to the set. And we assumed that an element x is not changed by being counted in the set (is not affected, for example, by being placed in the set, or coming into contact with some other element within the set)' (Nirenberg and Nirenberg 2018).

The basic characteristic of the elements of the set, the authors state in another work (Nirenberg and Nirenberg 2011, 606), is that they must be '*apathés,*' inert, which means that nothing happens to elements *x* or *y* when placed together: there is 'no change in identity' (Nirenberg and Nirenberg 2011, 607; also see 2018, 3, 8). The examples that the authors provide to contradict apathy are, among others, the chemical elements, which combine and transform when in contact (2011, 607).

According to Hardy (1940, 46), abstraction is exactly what characterizes mathematics. Numbers are not part of the world as we know it:

> when we assert that $2 + 3 = 5$, we are asserting a relation between three groups of 'things'; and these 'things' are not apples or pennies ... The meaning of the statement is entirely independent of the individualities of the member of the groups. All mathematical 'objects' or 'entities' or 'relations', such as '2', '3', '5', '+', or '=', and all mathematical propositions in which they occur are completely general in the sense of completely abstract. (see also 70–71)

The basis of the apathy pointed by the Nirembergs regarding the components of mathematical set is the principle of identity, the same cited above by Clastres: for every x, $x = x$. All any counting system requires is a first number, which may be zero or 1, and that the next number can be stated, 'such-and-such a number + 1.' What is not taken into account is that, in contrast to the stones that serve as the base for various counting systems (hence Borges's critique being based precisely on stones), '[o]ther things in the world are more difficult to count, whether because they are more subject to "becoming," as the philosophers would say, that is, to change in the act of counting, so that it becomes difficult to speak of the 1; or because they are subject to transformative interaction when brought together with another, so that we cannot speak of the $1 + 1$' (Nirenberg and Nirenberg 2018, 5). The authors also assert, in the other article mentioned, that the axioms of set theory imply that '[a]ny rigorous attempt to base an ontology upon them will entail such a drastic loss of life and experience that the result can never amount to an ontology in any humanly meaningful sense' (2011, 586).

The centrality of the unit for western mathematical thought leads the authors, along the same path as Lévy-Bruhl, to Christian monotheism: 'Note how math works here for monotheism: by modelling divine creation on the eternal identity of the number one and the repetitive move of the $1 + 1$, the multiplicity of the cosmos is created, while the unity of Being is maintained' (Nirenberg and Nirenberg 2018, 4).[19]

The blue stones and their non-apathetic behaviour lead me, before returning to the Wari', to two ethnographic contexts that are very distinct but, we could say, equally blue. One of them is the writing of the Tagalog of the Philippines in the seventeenth century, analysed in the monograph of Vicent Rafael. On arriving there and noting that the natives had a form of writing (probably originally from Java, Bali and Sumatra, for their part descendent from the writings of southern India, derived from the Brahmin), the Jesuit missionaries opted to use it for catechism. However, there was an aspect of this writing that they did not recognize and that eventually led them to abandon it in favour of Spanish.

While for the Spanish Catholic missionaries, writing existed independently of the speaker, for the Tagalog writing was completely subsumed to reading, which had an aleatory character where the same characters were associated with different sounds and meanings, depending on the reader. The term *baybayin*, designating writing, associated reading with floating 'over a stream of sounds elicited by the characters' (Rafael 1993, 49).

The idea of a fluctuating writing takes us to the second ethnographic context that of the Yaminawa (a Pano people of Acre, Brazil) who, though traditionally

---

[19]Simon Schaffer (personal communication, 2019) observes that, above all, this tale thematizes an epistemological/ontological shock, since the Scottish professor of logic epitomizes all those who believe that just one universe exists, functioning equally for everyone. According to him, the point is to know who is the Scottish logician in Wari' history: the rubber bosses, the evangelical missionaries or the intercultural teachers? My answer would be: all three.

agraphic, exhibited a similar instability in the meaning of the words uttered by shamans, which Townsley (1993, 460; see too Carneiro da Cunha 1998: 12–13) calls 'twisted words.' These were words from everyday language applied to very different objects – a fish might be called peccary, for example – a way found by the shaman to solve the dilemma of enunciating distinct perspectives in the same narrative (see Overing 1990, 610 for analogies with the Piaroa, and Lagrou 2007, 138 on transformations of form among the Kaxinawa/Huni Kuin, another Pano people).

The idea of an oscillating and indefinite world allows me to return to the Wari'. Until the solidification of Christian experience after a revival in 2001, two classificatory categories organized their entire lived world: *wari'*, which signifies us, people, predator, and *karawa,* which signifies animal, food, prey, and includes enemies, *wijam*, both people from other indigenous peoples and whites. These were not fixed categories, though, but positions, which could be occupied by different kinds of beings alternately. As I mentioned at the beginning of this article, either the Wari' saw themselves in the predator position and, therefore, human, or in the prey position, since this was how they were seen by animals, who saw themselves as humans (see Vilaça 2002, 2005, 2010, 2017). In sum, the distribution of beings in categories depended on the perspective of themselves and others and on the relational configuration in which they saw themselves inserted. In other words, even if we try to encounter analogies between *karawa* and our notion of nature, as some authors might have suggested over the course of an intense anthropological debate on the theme (Ellen 1996, 103, 113; Ellen and Fukui 1996, 10), it is essential to recognize that the core of its definition was its positional character, making *wari'* and *karawa* sets with variable contents and implying the idea of a highly unstable nature characteristic of perspectivism and, therefore, distinct from the idea that founds the scientific view – at least the school version presented to the Wari' (for similar critiques, see Strathern 1980 and Howell 1996, 135–140).

In the 1980s and 1990s, prior to the Christian revival, the shamans told me that they never enjoyed success on hunting trips and that their wives criticized them for the fact. The problem was that they looked at a deer and got ready to shoot, but it would suddenly turn into a paca, and then an armadillo and, sometimes, it appeared like a person. They were unable to have a fixed image precisely because these animals changed form. And not only this. The shamans also knew that an animal like the paca, for example, was, in the eyes of the jaguars, a papaya, or that maize beer was, for the latter animals, blood. There was, therefore, no stable universe of objects that formed sets capable of being identified or counted. Potentially, nothing was self-identical and, returning to the terms of the Nirenbergs, the elements of the *wari'* and *karawa* sets transformed constantly when in contact with each other.

This does not mean that there were no zones of stability in the Wari' world, since this was precisely what they sought to produce in their day-to-day lives, transforming others, including new-born children, into kin – that is, people who shared the same perspective and sustained a common world. This was, however, a derived apprehension considered limited by the Wari', since the gaze of shamans had epistemological privilege over other people: shamans are those with 'free eyes,' while the eyes of others are tied up (see Kopenawa and Albert 2013, on the Yanomami).

We do not need to turn to shamanism. The problems posed by relational and variable elements for set theory are explicit in everyday questions in the class-room. In a maths class on the Intercultural Teaching Degree course that I watched in April 2015, specifically on set theory, the teacher, based on the eth-nicity of his students (around 10 students), drew on the board a circle that he called a Tupari set and stated: 'Fernando is not a member of the Tupari set.' Immediately, a Suruí student recalled the children of mixed couples and asked a Tupari student if the patrilineality rule applied for them. The Suruí added that now they do the same as the whites; as his wife is Campé and he is Suruí, his daughter was registered as Campé Suruí. As the lesson continued, explaining the notion of membership, the teacher concluded: 'how can it be said that there is a Zoró who isn't a Zoró? That would be absurd!' We know, however, that this is precisely the case of Amazonian identities, since being Zoró is not a universal fact, or a reality, but one perspective among others, which must necessarily be associated with a person or group.

I wish to conclude with the fact that the Wari' opted to maintain the term 'alone' for the unit while adopting Portuguese nouns for the other numerical terms. Like diverse Christian translations among the Wari' (see Vilaça 2016), comparable to the 'twisted words' of the Yaminawa shamans, or even the port-manteau words used by Lewis Carroll in his *Alice,* allowing transitions between distinct ontologies (see Vilaça 2018), the 'alone' seems to me to have the same function in the school universe. It takes them from the 'universe of precision to the world of the more or less,' to invert an expression of Koyré (1980), and maintains the indefinition of elements as a constitutive part of their perspective universe.[20]

When Borges's professor of logic sought to find some pattern that would allow him to conceive a model for the multiplication of the blue stones, he dis-covered that the only clear principle was that a stone could not multiply when it was alone, isolated from the others: 'It took little experimenting to show that one of the discs, isolated from the others, could not multiply or disappear.'

---

[20]For a critique of Koyré's conclusions concerning the disinterest in precision in ancient Greek science, see Lloyd (1987, Chapter 5). I should make it clear here that I am not claiming that the Wari' chose to maintain the term for 'alone' as a form of conscious resistance to the imposition of the unit by Christians and by school mathemat-ics. My apprehension of this phenomenon, as well as that of the Christian translations as 'twisted words' or 'portals' between ontologies, relates to the movements of differentiation described by Wagner (2010), which are produced 'automatically' or unconsciously as a reaction to the conventionalization that comprises an intrin-sic part of this symbolic dialectics (my thanks to Stephen Hugh-Jones for obliging me to make this question explicit).

The isolation of the unit implies precisely the paralysation of the system evoked by Lévi-Strauss (1995), the end of the capacity of self-reproduction of the blue stones. The danger of the unit made explicit by the Barasana resides, then, not in its indefinition, which makes it open to the other, but precisely in its association with our 1.

Disclosure statement

No potential conflict of interest was reported by the author(s).

Funding

This work was supported by Conselho Nacional de Desenvolvimento Científico e Tecnológico: (CNPq) [Bolsa Pesquisador 2018] and by FAPERJ (Cientista do Nosso Estado 2018).

References

Almeida, Mauro. 2015 « Matemática concreta ». *Sociologia & Antropologia* 5 (3): 725–744.

Ascher, Marcia. 1991. *Ethnomathematics. A Multicultural View of Mathematical Ideas.* Pacific Grove: Brooks/Cole Publ. Co.

Ascher, Marcia. 2002. *Mathematics Elsewhere. An Exploration of Ideas Across Cultures.* Princeton-Oxford: Princeton University Press.

Ascher, Marcia, and Robert Ascher. 1969. "Code of Ancient Peruvian Knotted Cords (Quipus)." *Nature* 222: 529–533.

Biersack, Aletta. 1982. "The Logic of Misplaced Concreteness: Paiela Body Counting and the Nature of the Primitive Mind." *American Anthropologist, (N.S.)* 84 (4): 811–829.

Borges. 1998. "Tigres azules." In *La memoria de Shakespeare*. Madrid: Alianza Editorial. S.A. English version quoted: "Blue tigers". *The Independent*. Sunday, 27 December 1998.

Carneiro da Cunha, M. 1998. "Pontos de vista sobre a floresta Amazônica: xamanismo e tradução". *Mana Estudos de Antropología Social* 4: 7–22.

Carrier, Achsah. 1981. "Counting and Calculation on Ponam Island." *Journal of the Polynesian Society* 90 (4): 465–479.

Caux, Camila. 2015. "O riso indiscreto: couvade e abertura corporal entre os Araweté". PhD diss., PPGAS/Museu Nacional/UFRJ.

Caveing, Maurice. 1990. "Quelques précautions dans l'emploi de l'idée de nombre." *L'Homme* 30 (116): 151–157.

Clastres, P. 1989. "Of the One Without the Many." In *Society Against the State. Essays in Political Anthropology*, 169–176. New York: Zone Books.

Conant, Levi L. [1896] 2013. *Number Concept: Its Origin and Development*. Charleston: Nabu Press; Primary Source ed. Edition.

Crump, Thomas. 1990. *The Anthropology of Numbers*. Cambridge: Cambridge University Press.

D'Ambrosio, Ubiratan. 1990. *Etnomatemática*. São Paulo: Editora Ática.

D'Ambrosio, Ubiratan. 1994. "Prefácio." In *Com quantos paus se faz uma canoa! A matemática na vida cotidiana e na experiência escolar indígena*, edited by Mariana Ferreira, 7–11. Ministério da Educação e do Desporto e Comitê de Educação Escolar Indígena.

D'Ambrosio, Ubiratan. 2015. "Foreword." In *Mapping Time, Space and the Body. Indigenous Knowledge and Mathematical Thinking in Brazil*, edited by Mariana Leal Ferreira. Rotterdam: Sense Publ. ("New Directions in Mathematics and Science Education" 29).

Ellen, Roy. 1996. "The Cognitive Geometry of Nature: A Contextual Approach." In *Nature and Society. Anthropological Perspectives*, edited by Descola, Philippe and Gísli Pálsson. London: Routledge.

Ellen, Roy and Fukui, Katsuyoshi, eds. 1996. "Introduction". In *Redefining Nature. Ecology, Culture and Domestication*, 1–36. Oxford: Berg.

Everett, Daniel. 2005. "Cultural Constraints on Grammar and Cognition in Pirahã. Another Look at the Design Features of Human Language." *Current Anthropology* 46 (4): 621–646.

Everett, Daniel, and Barbara Kern. 1997. *Wari'. The Pacaas Novos Language of Western Brazil*. London, New York: Routledge.

Everett, Caleb and Keren Madora (ms). "Number as a Cognitive Technology: A Reevaluation of the Evidence".

Ferreira, Mariana Kawal Leal. 1998. *Madikauku: os dez dedos das mãos: matemática e povos indígenas no Brasil*. Brasília: MEC.

Ferreira, Mariana Kawal Leal. 2001. "People of my Side, People of the Other Side: Socionumerical Systems in Central Brazil." *Zdm. Mathematics Education* 33 (3): 89–94.

Ferreira, Mariana Kawal Leal. 2015. *Mapping Time, Space and the Body. Indigenous Knowledge and Mathematical Thinking in Brazil*. Rotterdam: Sense Publ. ("New Directions in Mathematics and Science Education" 29).

Frank, Michael, Daniel Everett, Evelina Fedorenko, and Edward Gibson. 2008. "Number as a Cognitive Technology: Evidence from Pirahã Language and Cognition." *Cognition* 108: 819–824.

Gonçalves, Marco Antonio. 2001. *Um mundo inacabado. Ação e criação em uma cosmologia amazônica. Etnografia Pirahã*. Rio de Janeiro: Editora UFRJ.

Gordon, Peter. 2004. "Numerical Cognition Without Words: Evidence from Amazonia." *Science* 306 (5695): 496–499.

Green, Diana. 1994. "O sistema numérico da língua Palikur." *Boletim do MPEG, Série Antropologia* 10 (2): 261–303. Belém: Museu Paraense Emílio Goeldi.

Green, Diana. 1997. "Diferenças entre termos numéricos em algumas línguas indígenas do Brasil." *Boletim do MPEG. Série Antropologia* 12 (2): 179–207. Belém: Museu Paraense Emílio Goeldi. https://silo.tips/download/diferenas-entre-termos-numericos-em-algumas-linguas-indigenas-do-brasil-1-diana

Hardy, G. H. 2018 [1940]. *A Mathematician's Apology*. Eastford, CT: Martino Fine Books.

Howell, Signe. 1996. « Nature in Culture or Culture in Nature ? Chewong Ideas of 'Humans' and Other Species », In *Nature and Society. Anthropological Perspectives*, edited by Philippe Descola and G. Pálsson, 127–144. London, Routledge.

Hugh-Jones, Stephen. 2012. "Escrita na pedra, escrita no papel." In *Rotas de criação e transformação. Narrativas de origem dos povos indígenas do rio Negro*, edited by Geraldo Andrello, 138–167. FOIRN/ISA.

Kopenawa, Davi, and Bruce Albert. 2013. *The Falling Sky. Words of a Yanomami Shaman.* Translated by N. Elliott and A. Dundy. Cambridge: Harvard University Press.

Koyré, Alexandre. [1943 and 1948] 1980. *Galileu e Platão e Do mundo do "mais ou menos" ao universo da precisão.* Translated by M. T. B. Curado, rev José Trindade Santos. Lisbon: Gradiva.

Lagrou, Els. 2007. *A fluidez da forma: arte, alteridade e agência em uma sociedade amazônica (Kaxinawa, Acre).* Rio de Janeiro: Topbooks. Ppgsa/Ifcs, Capes.

Lancy, David. 1981. "The Indigenous Mathematics Project: An Overview." *Educational Studies in Mathematics* 12 (4): 445–453.

Lancy, David, and Andrew Strathern. 1981. "'Making Twos': Pairing as an Alternative to the Taxonomic Mode of Representation." *American Anthropologist*, New Series 83 (4): 773–795.

Lave, Jean. 1988. *Cognition in Practice. Mind, Mathematics and Culture in Everyday Life.* Cambridge: Cambridge University Press.

Leite, Kécio2013. *Teia de conhecimentos interculturais.* Departamento de Educação Intercultural, Campus de Ji-Paraná, Universidade Federal de Rondônia (UNIR): 119-120.

Levinson, Stephen. 2005. "Comments on EVERETT, Daniel, Cultural Constraints on Grammar and Cognition in Pirahã. Another Look at the Design Features of Human Language." *Current Anthropology* 46 (4): 637–638.

Lévi-Strauss, Claude. 1963 [1958]. *Structural Anthropology.* New York: Basic Books.

Lévi-Strauss, Claude. 1978. *The Origin of Table Manners.* New York: Harper & Row.

Lévi-Strauss, Claude. 1981. *The Naked Man.* New York: Harper & Row.

Lévi-Strauss, Claude. 1983. *Structural Anthropology*, Vol. 2. Translated by Monique Layton. Chicago. The University of Chicago Press.

Lévi-Strauss, Claude. 1995. *The Story of Lynx.* Chicago: The University of Chicago Press.

Lévy-Bruhl, Lucien. 1985. *How Natives Think [Les Fonctions mentales dans les sociétés inférieures].* Translated by Lilian A. Clare. Princeton, Princeton University Press.

Lima, Tânia S. 1996. "O dois e seu múltiplo : reflexões sobre o perspectivismo em uma cosmologia tupi." *Mana. Estudos de antropologia social* 2 (2): 21–47.

Lima, Tânia S. 1999. "The Two and Its Many: Reflections on Perspectivism in a Tupi Cosmology." *Ethnos* 64 (1): 107–131.

Lima, Tânia S. 2008. "Uma história do dois, do uno e do terceiro." In *Lévi-Strauss. Leituras brasileiras*, edited by Ruben Caixeta de Queiroz and Renarde Freire Nobre, 209–263. Belo Horizonte: Ed. Ufmg.

Lizot, Jacques. 2004. *Diccionario enciclopédico de la lengua yānomāmi.* Caracas: Vicariato Apostólico de Puerto Ayacucho.

Lloyd, Geoffrey. 1987. *The Revolutions of Wisdom. Studies in the Claims and Practice of Ancient Greek Science.* Berkeley, Los Angeles, London: University of California Press.

Medeiros, Nádia, and Maxwell Miranda. 2009. "Análise preliminar dos quantificadores numéricos e não numéricos em Krenák (família Botocudo)." In *Anais – VI Congresso Internacional da Abralin.* Vol. 2, edited by Dermeval da Hora. João Pessoa: Ideia.

Mimica, Jadran. 1988. *Intimations of Infinity. The Cultural Meanings of Iqwaye Counting and Number System.* Afterword by Roy Wagner. Oxford, New Your, Hamburg: Berg.

Nirenberg, David, and Ricardo Nirenberg. 2011. "Badiou's Number: A Critique of Mathematics as Ontology." *Critical Inquiry* 37: 583–614.

Nirenberg, David, and Ricardo Nirenberg. 2018. "Knowledge from Pebbles. What Can Be Counted and What Cannot." *Know* 2 (1): 1–13.

Nykl, A. R. 1926. "The Quinary-Vigesimal System of Counting in Europe, Asia, and America." *Language* 2 (3): 165–173.

OroAt, Zebedeu. 2013. "Contagem do povo OroAt." In *Teia de conhecimentos interculturais*, edited by Kécio Leite, 171–172. Ji-Paraná: Universidade Federal de Rondônia.

OroNao', Mauricio. 2013. "Forma tradicional de contagem do povo OroNao." In *Teia de conhecimentos interculturais*, edited by Kécio Leite, 119–120. Ji-Paraná: Universidade Federal de Rondônia.

Ororam Xijein, Arão. n.d. *Experiências de ensino e pesquisa em ciências, meio ambiente e etnomatemática na licenciatura intercultural.*

Orowaje Cao, Wem Cacami. 2015. *Saberes matemáticos do povo Cao OroWaje.* Trabalho de Conclusão de Curso apresentado ao Departamento de Educação Intercultural da UNIR como requisito para a obtenção do título de licenciado em Educação Básica Intercultural, na Área de Concentração em Ciências da Natureza e Matemática Intercultural.

Overing, Joanna. 1990. "The Shaman as a Maker of Worlds. Nelson Goodman in Amazonia." *Man (n.s)* 25: 601–619.

Pica, Pierre, Cathy Lemer, Véronique Izard, and Stanislas Dehaene. 2004. "Exact and Approximate Arithmetic in an Amazonian Indigene Group." *Science* 306 (15): 499–503.

Robbins, Joel. 2004. *Becoming Sinners. Christianity + Moral Torment in a Papua New Guinea Society.* Berkeley, Los Angeles, London: University of California Press.

Seidengart, J. 2006. Infini. "Science classique". Dictionaire d'histoire et philosophie des sciences, edited by D. Lecourt et al., 605–611. Paris: PUF.

Silva, Vanilda Alves da. 2006. *Noções de Contagens e Medidas Utilizadas pelos Guarani na Reserva Indígena de Dourados – Um Estudo Etnomatemático.* Dissertação de Mestrado, UFMS, Campo Grande.

Strathern, M. 1980. "No Nature, No Culture: The Hagen Case." In *Nature, Culture and Gender*, edited by C. MacCormack and M. Strathern, 174–222. Cambridge: Cambridge University Press.

Strathern, M. 1992. "Qualified Value: The Perspective of Gift Exchange." In *Barter, Exchange and Value. An Anthropological Approach*, edited by Caroline Humphrey and Stephen Hugh-Jones, 169–191. Cambridge: Cambridge University Press.

Tabak, John. 2011. Mathematics and the Laws of Nature: Developing the Language of Science, Revised ed.

Tenório, Higino, José Ramos, and Flora Cabalzar. 2004. *Keore. Utapinopona saiña hoa bauaneriputi. (Tuyuka Maths: A Guide for Continuous Research)*, 48 p. São Gabriel da Cachoeira: AEITU.

Townsley, Graham. 1993. "Song Paths. The Ways and Means of Shamanic Knowledge." *L'Homme* 33 (2–4): 449–468.

Urton, Gary, and Primitivo Llanos. 1997. *The Social Life of Numbers: A Quechua Ontology of Numbers and Philosophy of Arithmetic.* Austin: University of Texas Press.

Valdenilson, Takwyiti Hompryti, and Aline da Silva Lima. 2017. (ms) *Entender o saber matemético do povo Pàrkatêjê.*

Vernant, Jean-Pierre. 1972. *As origens do pensamento grego.* São Paulo: Difusão Européia do Livro.

Verran, Helen. 2001. *Science and an African Logic.* Chicago: University of Chicago Press.

Vilaça, Aparecida. 2002. "Making Kin out of Others in Amazonia." *Journal of the Royal Anthropological Institute* 8 (2): 347–365.

Vilaça, Aparecida. 2005. "Chronically Unstable Bodies: Reflections on Amazonian Corporalities." *Journal of the Royal Anthropological Institute* 11 (3): 445–464.

Vilaça, Aparecida. 2009. "Bodies in Perspective: a Critique of the Embodiment Paradigm from the Point of View of Amazonian Ethnography." In *Social Bodies*, edited by Helen Lambert and Maryon McDonald, 129–147. New York and Oxford: Berghahn Books.

Vilaça, Aparecida. 2010. *Strange Enemies. Indigenous Agency and Scenes of Encounters in Amazonia. Transl. by David Rodgers*. Durham: Duke University Press.

Vilaça, Aparecida. 2016. *Praying and Preying. Christianity in Indigenous Amazonia.* Translated by David Rodgers. Oakland: University of California Press.

Vilaça, Aparecida. 2017 [1992]. *Comendo como gente. Formas do canibalismo wari' (Pakaa Nova)*. Rio de Janeiro: Mauad X.

Vilaça, Aparecida. 2018. "Le diable et la vie secrète des nombres". *L'Homme* 225: 149–170. [Translated into English as "The devil and the hidden life of numbers. Translations and transformations in Amazonia. The Inaugural Claude Lévi-Strauss lecture." *HAU: Journal of Ethnographic Theory 8 (1/2): 6–19,* and into Portuguese as "O diabo e a vida secreta dos números. Traduções e transformações na Amazonia". *Mana. Estudos de Antropologia Social* 24 (3), 2018].

Vilaça, Aparecida. 2019. "Inventing Nature. Christianity and Science in Indigenous Amazonia." *Hau. Journal of Ethnographic Theory* 9 (1): 44–57.

Viveiros de Castro, Eduardo. 1996. "Os pronomes cosmológicos e o perspectivismo ameríndio." *Mana. Estudos de antropologia social* 2 (2): 115–143.

Viveiros de Castro, Eduardo. 1998. "Cosmological deixis and Amerindian perspectivism." *Journal of the Royal Anthropological Institute* 4 (3): 469–488.

Viveiros de Castro, Eduardo. 2000. "Atualização e contra-efetuação do virtual na socialidade amazônica: o processo de parentesco." *Ilha* 2 (1): 5–46.

Viveiros de Castro, Eduardo. 2001. "GUT Feelings About Amazonia: Potential Affinity and the Construction of Sociality." In *Beyond the Visible and the Material: The Amerindianization of Society in the Work of Peter Rivière*, edited by Laural Rival and Neil Whitehead, 19–43. Oxford: University Press.

Viveiros de Castro, Eduardo. 2007. "The Crystal Forest: Notes on the Ontology of Amazonian Spirits." *Inner Asia* 9: 153–172.

Viveiros de Castro, Eduardo. 2012. "Radical dualism. A meta-phantasy on the square root of dual organizations, or A savage homage to Lévi-Strauss". In *documenta* (13), *100 Notes – 100 Thoughts / 100 Notizen - 100 Gedanken*, no. 056, 5–22. Berlin: Hatje Cantz, 2012. Published in conjunction with the Documenta 13 exhibition in Kassel, Germany.

Viveiros de Castro, Eduardo. 2013. "Cannibal Metaphysics: Amerindian Perspectivism." *Radical Philosophy* 182. (November/December 2013).

Wagner, Roy. 2010 [1975]. *A invenção da cultura*. São Paulo: Cosac Naify.

Wierzbicka, Anna. 2005. "*Comments* on EVERETT, Daniel, Cultural Constraints on Grammar and Cognition in Pirahã. Another Look at the Design Features of Human Language." *Current Anthropology* 46 (4): 641.

As perceived, not as known: digital enquiry and the art of intelligence*

Willard McCarty

ἐὰν μὴ ἔλπηται ἀνέλπιστον οὐκ ἐξευρήσει, ἀνεξερεύνητον ἐὸν καὶ ἄπορον. ('If you do not expect the unexpected, you will not find it; for it is hard to be sought out and difficult')
   Heraclitus (DK 18).

1. Introduction

What happens when you use a digital machine to explore a serious question and somehow, during the interactions that follow, find yourself with a result that is or promises to be significant? What role might the rule-bound machine have? For the last several decades, work in the cognitive sciences and in anthropology has variously argued for embodied, extended or distributed systems of doing thinking with the world's affordances; in his influential theory, Gibson compared such thinking with the life of an animal in its ecological niche.[1] What part might the switching circuits of the digital machine be playing in our niche, and how do they play it?

   Though these are not particularly easy questions, their aim is relatively modest and has been pursued for decades. But my reason for posing them is only preparatory, to elicit sufficient knowledge about the machine that will allow me to shift attention from the usual subject–object perspective to the cognitive betweenness of the human–machine relation. What I mean by this will become clear, or at least clear enough for me then to advance the question I really want to ask: how might the combinatorial potential of these switching circuits be deployed, not merely to vend information (what we usually do with them) or mimic some aspects of human behaviour (the dominant aim of AI),

*For inspiration, unwavering encouragement and invaluable resistance I am esp. indebted to G. E. R. Lloyd (*nisi tu*). I am also greatly indebted to my fellow participants in SFSP II, esp. Evelyn Fox Keller, Francesca Rochberg, Marilyn Strathern and Aparecida Vilaça; and to those who served as readers: Brad Inwood first of all, also Bob Amsler, Jan Christoph Meister, Francesca Rochberg, Tim Smithers and Manfred Thaller. Whatever muddle and error that remain, I acknowledge mine.
[1]Gibson 2015/1979, 120–1. For the 'extended mind' hypothesis, see e.g. Jianhui (2019), Menary (2010) and Lenoir (2007); for the related 'distributed cognitive systems', e.g. Perry (2017), Giere (2002) and Hutchins (1995); for 'embodied cognition', Wilson and Foglia (2017). In anthropology, see Ingold (2010) and Gell (1998, Chapter 9). The phrase 'doing thinking' is Mahoney's (2011, 87).

but to meet an enquirer's thought-processes half-way, with all the alterity of a very differently constructed and surprising 'intelligence'?[2]

Full disclosure: I start from the simplifying assumptions that the user is an allegorical Euro-American Everyman and that the machine in question is the current model, designed to match Everyman's needs if not awaken or even create them. My wager is that these parochial assumptions will allow me economically to reach to something that is not so culturally limited as I am apt to be, indeed to something fundamental. To make this reach, I call for help on three analogies – to ordinary conversation, laboratory research and traditional divination.[3] My belief is that by drawing them together 'some understanding is possible, however incomplete, provisional, and revisable that is' (Lloyd 2019, 38).

Background to my bridge-building attempt is an interdisciplinary effort over a number of years to bring long-established disciplines within reach of the nascent digital humanities in order to suggest new possibilities of action for a more broadly conceived and intellectually serious discipline. Hence, most obviously, the length of the study, its bibliographic exuberance, the necessarily brief treatment of several areas of research and the not inconsiderable risks.

First, in Sections 2 and 3, I scrutinize those switching circuits and what can or might be done with them. Then, in Sections 4–6, come the analogies. Conversation leads off. I consider work in sociolinguistics and in the evolution and development of intelligence to bring out aspects of 'talk-in-interaction' (Schegloff's term), especially important for modelling human–machine relations.[4] Laboratory science comes next. Studies in its history and cognitive psychology illumine the obvious parallel to computing in the investigative, epistemic use of physical instruments. Finally divination. Its primaeval, world-wide and highly diverse practices likewise offer a way of seeking answers to questions by physical means. But crucially these practices add the badly needed reach beyond the parochial, to the historical and cross-cultural depth of an intensely studied anthropological phenomenon. Divination thus challenges and enriches current, differently technological means. I consider these analogies one by one, then I align them, in Section 7, to sketch out a possible common ground. Section 8 concludes.

2. The machine

The standard account of enquiry by computer centres on *modelling*, that is, recursively making, probing and changing a digital likeness of something.[5]

---

[2]From the perspective of computer science, see Winograd (2006). On alterity, see Taussig (1993), Castoriadis (1987/1975, 178), Levinas (1999/1995), cf. Auerbach (2003/1953).

[3]Other analogies than the ones I have chosen would be fruitful to explore but could not be included for want of space, e.g. weaving and related practices, with connections to studies in symmetry, art history and anthropology, among several other fields.

[4]Turing (1950), Kay (1969) and Hutchins (1987). See also Neale and Carroll (1997).

[5]See McCarty (2019b), Ciula et al. (2018), McCarty (2014/2005) and Morgan (2012). For simulation, in relation to modelling, see McCarty (2019a).

I prefer 'modelling' to 'model' because the verbal noun places strong emphasis on an open-ended, enquirer-inclusive process and the ongoing manipulation of software with which it is done. 'Model' suggests a stable, theory-like object, something that will suffice. Modelling, closer to metaphor in its instability and emphasis on difference,[6] creates a dynamic three-way relationship of object, model and modeller and so is interpretative at every step. It happens in three normally recursive phases: translating the object of interest into binary data and one's questions into software; manipulation of the data by hardware; and interacting with the machine to make sense of the results. This essay is concerned primarily with the last and least well-understood phase, when scholar and machine get close, and something involving both takes place. Rather than focus on interaction with an autonomous AI, such as may be, I prefer to think towards a *relational* intelligence, created in the asymptotic and asymmetrical convergence of human mind and computational affordance.[7]

To get there, I begin with Edsger Dijkstra's good advice: to shift attention from the user-friendly interface back towards the hardware, bypassing the many layers of accommodating software (1986: 48). I stop, short of the electronics, at the 'level of abstraction' (Colburn and Shute 2007) at which the machine's unaccommodated and unaccommodating contributions to such relational intelligence become visible. Of course, we need these layers of systems software in order to render the power of digital computation usable for the kinds of problems that concern us. But at the same time, their net effect, indeed their purpose, is to hide what the machine actually does, and so to conceal any 'implementation trace' or imprint the hardware might have on what the machine does.[8] Although whatever may be said to be *of* the machine must also be *of* its human maker, the machine as a work of art is paradoxically capable of deeply estranging effects. Polish artist Bruno Schulz once described the role of art as 'a probe sunk into the nameless' (1998, 369–70). I argue for the machine's potential to play a like role in relation to ourselves. It can indeed *surprise* us (who know not ourselves terribly well). By means of it, *we* can surprise us in surprising ways.

In his 1950 provocation on machine intelligence (Gandy 1996), Alan Turing wrote that if we are caught in error by the machine, or if we take the occasion to surprise our conscious selves, 'no credit on the machine' accrues, but when by its relentless thoroughness, it illumines that which our expectations prevent us from seeing, we glimpse a different vista of mind (Turing 1950, 450–1). Why is

---

[6] I echo Evelyn Fox Keller's insights in Keller (2020); see also the commentary on it in *Interdisciplinary Science Reviews* 45.3.

[7] On human relations (which is where we begin), see Strathern (2020); with machines, Goffey (2008).

[8] There is an argument here, for which see Turner and Angius (2020, 50) (pdf version). I side with Colburn and Shute (2007); see also Kittler (2013, Chapters 15–16), Kirschenbaum (2008), Evens (2015), Chesher (1997); also Parhami (2003) for a specific example.

this so? Expectations are not the problem, for without them is no surprise, no hinterland at the liminal borders where the new can be perceived or created.[9]

Another way of dealing with the human–machine relation is to take it in the same way Evelyn Fox Keller suggests we consider the question of whether molecular biological constructs are alive: that it is not so much a philosophical as 'a historical question, answerable only in terms of the categories by which we as human actors choose to abide, the differences that we as human actors choose to honor, and not in either logical, scientific, or technical terms' (2002, 294). The move that I will be recommending is to constrain choice by the specific determinants of the machine and the differences these make, then to work out what sort of a cognitive relation is possible.

Before that, however, we must remain focused on the machine at that level of abstraction, to consider its hardware-determined *nómos*.

3. Multiple choice, revision, complexity and combinatorics

To strip away familiarity, we begin with digital hardware and logico-mathematical software (footnote 7). Hardware provides software with a remarkably frugal set of instructions. Roughly speaking, these allow a programme to carry out primitive arithmetic and logical operations on binary data, store the results in memory, move data from one location to another, change the order in which operations are performed and handle input and output.[10] That's all. But perhaps the starkest and for the scholar, the most immediate lesson is delivered by preparing an object of study for processing, i.e. translating it for input according to the machine's Procrustean all-or-nothing template (cf. McCarty 2019c, 153). This template requires that everything to be processed, in every detail, must be rendered according to what I call the axioms of digitization: *complete explicitness and absolute consistency of representation*. Doing that should effectively shatter 'the idealised and serene process that we imagine' or comfortably assume computing to be (Hasslacher 1995, 391).

The question then to be asked hardly needs spelling out: what contributions to a relational intelligence can possibly emerge under those conditions? In what terms can we conceive a relation between ourselves and such a device? But a better question is Marilyn Strathern's: 'the kind of connection one might conceive between entities that are made and reproduced in different ways – have different origins in that sense – but which work together' (2005, 37).

---

[9]See e.g. Kolmogorov's notebook entry in 1943 on the 'fine layer between the "trivial" and the impossible' where mathematical creativity takes place (Shiryaev 2000, 50); Huron on listeners coming to expect the unexpected in modernist music (2007, 333 and *passim*; cf. Merton 1968, Chapter 13); also the literature on creativity, e.g. Kaufman and Sternberg (2010) and Kirsh (2014); Hyde on the cultural history of the trickster (1998).

[10]Thanks to Manfred Thaller for this point (private e-mail, June 10, 2020), given detailed exposition by Hennessy and Patterson (2012, Appendix A), and broader context in relation to combinatorics by Fodor (1983, 29). I consider only the so-called von Neumann architecture of hardware here.

In his last lecture, John von Neumann laid down the implicit challenge in physico-mathematical terms. Based on 'logics and mathematics in the central nervous system … viewed as languages', he concluded that 'whatever the system may be' by which the brain communicates, 'it cannot fail to differ considerably from what we consciously and explicitly consider as mathematics' (1958, 82). This difference remains 'one of the outstanding unsolved problems of theoretical neuroscience'.[11] It suggests a trajectory for our enquiry that, unlike much current work in artificial intelligence, strives to highlight rather than downplay or ignore fundamental differences (cf. Watson 2019). Brought to light, such differences would provide a cogent basis from which to imagine and develop a relation that entails neither a distortion of the human nor a diminishing of the machine. What do we see when we get close enough to the machine to see these differences?

3.1 The semi-opaque 'black box'

The frugal vocabulary of hardware (its 'instruction set') does not give us much help in understanding what software can do, likewise the severely reductive constraints of its binary template. Understanding comes, rather, from the art of programming, that is, from the skilled making of things, step by step, with that vocabulary of instructions under their defining constraints. For my purposes, the most economical statement is provided by Herman Goldstine's and John von Neumann's 1947 report on what programming the machine might involve.[12] Their central insight was that programming the machine 'is not a static process of translation, but rather *the technique of providing a dynamic background to control the automatic evolution of a meaning*'.[13] This background can specify a linear, fixed sequence of operations to be followed, mimicking the simplest mechanical device. But the power special to the digital machine, they point out, is realized when software allows non-linear operations by exploiting two crucial abilities provided in hardware: first, to interrupt the sequence of instructions conditional on the outcome of prior events, such as results from computations or (especially significant in later developments) external triggers or 'interrupts'; second, to rewrite instructions during the course of programme execution.[14]

---

[11]Wells (2007, 1); cf. Israel and Millán Gasca (2009, 162–5).

[12]Goldstine and von Neumann (1947); cf. Goldstine (1972, 266–70). The machine they had in mind had not yet been built and was not operational until ca. 1952. See Knuth and Pardo (1976, 16–20), Aspray (1990, Chapters 3–4 and esp. 63–72), Campbell-Kelly (2011, 24–8); Gramelsberger (2011, 138–9), Chun (2011, 25, 191 n. 27), Israel and Millán Gasca (2009, 149). Hagen compares their flow-diagrammed scheme to a musical score (2006, 167–8).

[13]Goldstine and von Neumann (1947, 2, my emphasis). They do not stop to define 'meaning' but seem to have in mind something like significance for the problem at hand.

[14]Mark how close these non-linear operations are to what we do in conversation and in other uses of language (J. C. Meister, e-mail, December 27, 2019). The analogy to conversation is indeed quite strong.

When fully realized, that is, although the capabilities of hardware, the data to be worked on, the software instructions that might be executed and the results are all knowable, how these results are achieved is not, nor in general can we reliably predict what they will be. One could step the machine through each instruction one by one; the instructions could be printed out for inspection; but the complexity of the whole at speed would overwhelm any observer. Note especially 'complexity' and 'at speed'.

3.2. Complex, chaotic, random

When our purpose in using such a machine is to match whatever it models closely (e.g. tomorrow's weather), reliable prediction is primary; surprises are likely to indicate deficiencies of information, errors in software or the need to adjust or rethink the model. But when the aim is to work out what something might be were conditions other than they are, to help imagine what we do not know (McGann's phrase), then software can be tuned, as hardware makes possible, to admit surprise, to edge somewhat closer to disordering or innovating randomness 'at *the edge of chaos* ... the constantly shifting battle zone between stagnation and anarchy, the one place where a complex system can be spontaneous, adaptive, and alive'.[15] In consequence of this controlled tuning or relaxation of constraints, the system becomes capable of manifesting 'complexity' in the technical sense: in simplest terms, the behaviour characteristic of a non-linear system in which interaction among components predominates.[16] As a complex system, given the speed at which machines now operate, its actual operations are doubly unknowable, that is, unpredictable and irrecoverable after the fact. But the important point for the questions I am raising here is that precise knowledge of the sequence is irrelevant.

Ordinary online searching provides a familiar example. It is enacted by two agents: the search engine, whose strategies are defined and carried out by its algorithms, and the enquirer, whose words or phrases indicate or gesture towards something. Both query and strategy evolve as the enquirer adjusts the search and tries again. In very different ways and senses, both engine and enquirer learn, each from the response of the other. In addition, an engine (such as Google's) 'learns' from the queries and responses of all who have used and are using it and register their preferences silently by the choices they make.[17] Within calculated limits, uncertainties and surprises, once regarded as problems to be solved by supplying better information, become possible opportunities. For most of us doing research with the aid of online resources, the former goal (to converge on a right or optimal answer) has

---

[15]Waldrop (1992, 12); see also Chapter 6 and Langton (1992).
[16]Holland (2014) and Gallagher, Appenzeller, and Normile (1999). For self-organization, see esp. Keller (2008; 2009).
[17]Brin and Page (1998). For 'machine learning', see Watson (2019).

been overcome or at least seriously demoted in consequence of the literally exponential growth of data in volume, dynamics and variety. In this way, we are nudged if not pushed to go wide rather than deep and so to act in accord with the belief that 'the more integration between … [diverse] descriptions, the better is our understanding of the object identified by any of those descriptions' (Rorty 2004, 24).

In general, then, vast amounts of data and fast machines are capable of constituting and modelling phenomena described in the vocabulary of 'complex systems'.[18] These systems (such as the Internet or phenomena we view as systems, e.g. the weather, eddies in a stream, an ecological niche or social group) are characterized by 'chaotic', seemingly 'random' behaviour. In complexity theory, these terms are used neither in the ordinary nor metaphysical senses; rather they denote unpredictable behaviour observed to arise in deterministic systems that are very sensitive to small changes.[19] Such systems turn out to be where you may not expect them to be (e.g. the swirling water in your kitchen sink), once you look with complexity in mind.

For the outsider, this terminology throws up difficulties. The three key words – 'complex', 'chaotic', 'random' – are rarely explained. No one definition of any of them has achieved consensus. Usage differs considerably across the mathematical, physical and computational domains in which they are to be found.[20] Nevertheless, they are about something recognizable and important; their study has changed the natural, social and computational sciences radically and has had considerable influence more broadly. Allow me to suggest that historically the invention of the digital machine led to rapid growth of studies in what became the sciences of complexity by providing a matrix (as if made for the task) within which dynamic models of systems perceivable or discoverable as 'complex' could be constructed. To put the matter another way, it seems evident that the non-linear modelling power of the computer provided nigh irresistible inducement to look at old things in a new way, as *complex*, and to bring to light complexities previously unnoticed in the sciences – phenomena that are orderly but 'chaotic', paradoxically revealing structure and evading closure.

In the hands of its developers, the machine did more than that, however, as Peter Galison tells the story of its earliest uses: 'In the baldest possible form: the computer began as a "tool" – an object for the manipulation of machines, objects, and equations. But bit by bit (byte by byte), computer designers deconstructed the notion of a tool itself as the computer came to stand not for a tool, but nature itself' (1996, 156–7). The blurring he describes – or, one might say,

---

[18]For observations from the mid-nineteenth century, see Alligood, Sauer, and Yorke (1996: vi–vii) and Keller (2009). For 'chaos', see Bishop (2017, 3–6 (pdf)) and Smith (2007); for 'complexity', Holland (2014).

[19]Stewart (1989, 16f); Hao (1989, 1–4), Lorenz (1993); for the history of the idea, Keller (2008; 2009) Alligood, Sauer, and Yorke (1996: vi–viii).

[20]On the difficulty of definition, Hao (1989, 4) and Shermer (1995, 71f); on mathematical, physical and computational definitions of chaos, Smith (2007, 53–7, Chapters 4 and 6).

following Lloyd (1991) and Rochberg (2016), the recovery of 'nature' as an invented, culturally contingent way of seeing the world – is the further point. For this reason, I suspect, we find a qualifying language of *seeming* running through explanations in the empirical sciences of complexity: the language of what can only be known *from appearances*, the observable result of unknowable operations. I note the crucial importance of visualization (i.e. making visual by mathematical means) to the spread of these sciences; the lack of agreed definitions and diversity from discipline to discipline; the inclination to settle for 'a working or operational definition' of key terms (Hao 1989, 4); and the 'I know it when I see it' defence (Miller and Page 2007, 4). All these, I suggest, point to the uncertainty fundamental to a world seen as complex and to the modeller-builder's approximate, incomplete but progressive step-by-step depicting of how it works. Consider that a machine which to be useful requires us to engage in modelling, and which thus weakens the nature/culture dichotomy, frames a constructivist's perspective on everything.[21]

As with the molecular geneticist Jacques Monod's elegant scheme of the complex adaptive machinery of life (1972/1970), in which genetic necessity is animated by cosmic chance, all complex adaptive systems are in their way made possible by boxing in the randomness of the chaotic that animates them. But note: like 'chaos', 'random' eludes agreed-upon definition; it frustrates 'our pattern-finding instincts', throwing up a barrier to an intuition that has been shaped by our natal culture.[22] 'Random' is further refracted necessarily in its mirroring of the differing concepts of order it violates across particular needs, disciplines, historical periods and cultures. In ordinary usage, 'random' denotes the haphazard, the aimless, directionless and purposeless (*OED*; Keynes 1921, chapter 24). In computer science, it is, again, a term of seeming, qualified to mean the absence of any *discernible* relation between a number or object and any other in a digitally generated sample.[23] Hence, a provisional definition of 'random' might simply be the state of something beyond our ability to conceptualize, and 'randomization' the process of approaching or achieving that state.

We are left, then, with the dialectic of surprise and expectation, or in the language of complexity theory, at 'the edge of chaos', in 'an essential tension which keeps the ongoing dynamics on an indefinitely extended transient, far from equilibrium'.[24]

One further bit of mathematics remains.

---

[21]For a lucid account of computational worldmaking with 'complex adaptive systems', Lego-like brick by Lego-like brick, see Holland (1998), cf. Miller and Page (2007) and for the relevance of Lego, Cook and Bacharach (2017) and Dawson, Dupuis, and Wilson (2010).

[22]Seife (2014, 50). On randomness, see Beltrami (1999), Bennett (1998), Lorenz (1993), Brown and Keene (1957), Keynes (1921, chapter 24) and Poincaré (1914/1908, 64–90). Cf. Johnson (2015), Wagner (2012) and Dennett (1995). Note Thom's demand that we scrap the whole idea of randomness (1980: 1).

[23]Knuth (1998, 2); also L'Ecuyer (2004, 37) and both for the practical uses of random number generation.

[24]Langton (1986, 129) and all of *Physica D* 22 for context; cf. Waldrop (1992).

3.3. Combinatorics

To review: 'complex' characterizes phenomena considered as systems of inter-related parts in non-linear relation to each other; 'chaotic', a dynamical state complexity makes possible; 'random', the apparently disconnected relation of one datum to another. To these, I add 'combinatorics', or analysis of 'combinations of objects belonging to a finite set in accordance with certain constraints'.[25] It is the mathematics closest to what the enquirer does with the help of software (cf. von Neumann 1951, 16; Fodor 1983, 29). It is, Claude Berge explains, 'a matter of "configurations" which arise every time objects are distributed according to certain predetermined constraints'.[26] 'With combinatorics, one looks for their intrinsic properties, and studies transformations of one configuration into another, as well as "subconfigurations" of a given configuration' (Berge 1971/1968, 1–2). If you see serious, studious play at work here, you understand.

Again my question: how can behaviour we recognize as intelligent arise from or in relation with a device restricted to a radically frugal instruction set and restricting all input to complete explicitness and absolute consistency? We have unnecessary difficulty with this question, I think, because we expect theoretical depth as a precondition of significance and a sign of intelligent behaviour. We tend to look for explanation of computational results supposing that they are governed by fundamental law-like abstractions, but what we get, in the situations of complexity normal to the machine (Goldstine and von Neumann 1947, 2), is not like that: no law-like adherence, rather constraints which define the machine's recombinatorial potential. In any case, combinatorics in its simplicity and practicality rules out theoretical depth, hence the long-standing complaint of mathematicians.[27] But with the spread of digital computing and cross-cultural studies in the history of combinatorics and anthropology of mathematics (Pais 2013), denigrations have become ever less credible and ever more difficult to hear. Insofar as computing is concerned, attention needs to shift from explanation and search for meaning in the results to their utility, from the depth of theory to breadth and power of utility.

Consider combinatorial board games – draughts, chess and *go* (Chinese *weiqi*), for example.[28] Studies of these illumine how genuinely creative scholarly enquiry with the digital machine is possible. The board game gives us the

---

[25] *OED* s.v. See Wilson and Watkins (2013) and Lovász (2005); for computer science, Lovász, Shmoys, and Tardos (1995).

[26] For Berge, see https://bademian.wordpress.com/2012/10/06/a-tribute-to-claude-berge/ and Rota's Preface in Berge 1971/1968. Berge was a founding member of l'Ouvroir de Litterature Potentielle (Oulipo) and its most important mathematician; see Motte (1998/1986).

[27] Kleitman (2000, 124); cf. Berge (1971/1968) and Kung (1995). For its history, see Bréard (2015), Wilson and Watkins (2013), Netz (2009) and Biggs, Lloyd, and Wilson (1995). For combinatorial reasoning, DeTemple and Webb (2014); for the relation to ethnomathematics, Ascher (1991); for the relation to games, Conway (2001) and to divination, David (1962).

[28] Fairbairn (2007) and Finkel (2007 *passim*), Conway (2001), David (1962), cf. Nicolau (2009) and McCarty (2019, 154).

physical constraint of the board and the objects to be rearranged on it; in addition, it gives us a set of simple rules to be followed in rearranging them and a definition of a winning configuration. Complexity, sometimes profound, sometimes innovative, follows, as the histories of chess and *go* demonstrate. 'As with seeds', Holland observes, 'much comes from little' (1998, 1). To follow his metaphor, if the seed's genomic material is the board, game-pieces and rules, then the player is the epigenetic environment shaping how the possibilities of development play out (but cf. Keller 2020). Note that randomization, productive of surprise, is only sometimes explicit (e.g. in Snakes and Ladders and Monopoly) but is plausibly among the cognitive factors in a player's choice of moves in such games, given that these factors are unobservable and so may appear 'random' to observers, or indeed to the instinctual player.

The machine's combinatorics at speed offer a uniquely practical way of applying the mathematics to those phenomena that, in the words of the Manhattan Project historian David Hawkins, are 'too far from the course of ordinary terrestrial experience to be grasped immediately or easily', or, indeed, at all.[29] From that thought, it is a short step to prosthetic, tele- or microscopic metaphors pointing into a future (in the language of the later *Star Trek*) 'where no one has gone before'.

In sum, the mathematics of the machine shows us, for indefinitely many worlds constructible in its terms, a tunable, agile ability to defamiliarize what we thought we knew and how we knew it – indeed, to unsettle knowing, to knock us back to perceiving. It is at such moments, wondering how to grasp the situation, that the interdisciplinary and cross-cultural imperative points the way.

It's time, then, to tune in to those analogies and the immensely helpful worlds of scholarship they bring into focus. But I must ask for your patience. I make my case for their relevance at the end.

4. Conversation

Conversation, especially the ordinary kind, is a seemingly humble subject. But it is a rich and consequential area of study, relevant to the third phase of computing, when we cast it as interactive *languaging* with others and ask what comes of such talk. Languaging became a sociolinguistic specialism in the wake of J. L. Austin's work on 'performative utterances'[30] and the subsequent writings of Emmanuel Schegloff and others in the field that emerged from their empirical attention to 'talk in interaction', finding in it a site of coordinated and

---

[29]Hawkins (1946, 76); cf. https://www.atomicheritage.org/profile/david-hawkins (March 4, 2019).
[30]Austin (1962). See Wisdom, Austin, and Ayer (1946, 173ff) for the philosophical context; cf. Swain (2006) and Markoš, Švorcová and Lhotský (2017). On recent cognitive neuroscience of languaging, see Anderson (2014, Chapter 7).

contextually rich social action.[31] Hence, Conversation Analysis (CA). But however rich a lode, empirical observation of conversational exchange gets us only so far (Dale et al. 2014): the participants' frame of reference remains tacit,[32] and in what can be observed, Esther Goody notes, there is 'strong evidence for the patterning of interaction *on a level of which we are not aware* … communicative microstrategies which permit delicate adaptation to each other's responses … ' (1995, 22; cf. Pask 1961, 35). The 'recipient-designed' conversational moves discovered in CA point off-stage to more than can be transcribed, including but not limited to the conversational actors' physical response, e.g. in expressions or gestures (Turnbull 2003, 42; Anderson 2014, 256–7). Nevertheless, as I suggested earlier (footnote 14), the non-linear possibilities permitted by computational hardware and the common experiences of talk-in-interaction align. Can we pick out an originating imprint of ordinary human conversation in Goldstine's and von Neumann's conception of that 'dynamic background to control the automatic evolution of a meaning'?

CA, valuable as it is for my concerns here, does leave us wanting at the threshold. We need to stretch the notion of the interaction which it documents in order to capture the cognitive intimacy, moment-by-moment coordination, synchronization and performative modification that close analysis of conversation actually suggests.[33] Perhaps an analogy to musical performance in terms like 'resonance' and 'improvization', suggested in the mid-1970s for human–computer relations (Thompson 1974, 227) and subsequently for conversation,[34] is much closer to what happens, more *conversational*, more faithful to the 'mutual *tuning-in* relationship' or 'repertoire in action', than 'interaction'.[35] But, given that the word is everywhere in use, I recommend stretching rather than abandoning it, to include both improvization and resonance.

Coming from a very different direction, recent evolutionary research on the startling growth of hominid intelligence reinforces the creative role of conversation. This research shows that the evolution of primate cognition may well have been powered by languaging and other communicative actions in the negotiation of social encounters among gregarious hominids. Thus, the so-called Machiavellian Hypothesis that 'social intelligence' originated in efforts

---

[31]For early attention to conversation as a model of interactive behaviour, Reichman (Adar) 1986. For a history of conversation analysis (CA) in linguistics, see Fox et al. (2013); for CA overall, Sidnell and Stivers (2013) and Schiffrin, Tannen, and Hamilton (2001); in studies of intelligence, Goody (1995); in anthropology, Moerman (1988) and Clemente (2013); in cognitive and developmental psychology, Turnbull (2003), Edwards and Middleton (1987) and de Rosnay et al. (2014); improvization, Sawyer (2003).

[32]Pask (1961, 35). On Pask's cybernetic theory of conversation, see Pickering (2010), and on its relation to Pask's 'aesthetically potent environment' of the arts, Reichardt (1971).

[33]Hence, Schegloff's 'talk-in-interaction' (1991, 152 n1) and Laurel's reference to 'common ground' in the context of the theatricality of graphical interfaces (2014, 3–5); cf. te Molder and Potter (2005, 26–28), Streeck (1995), Drew (1995) and Newman-Norlund et al. (2009).

[34]Duranti (1997, 2015), Duranti and Burrell (2004), Duranti, Ochs, and Schieffelin (2012) and Sawyer 2003.

[35]On the tuning-in, Schütz (1951); repertoire in action, Faulkner and Becker (2009). Cf. Johnson-Laird (2002), Wagner (1996, 88), Winch (1959, 93f); music and interaction, Mukerji (1998), Star (1998); conversation as metaphor in improvisational music, Berliner (1994), Sudnow (2001), with Gibbs (2005, 75–77). Klerer suggests the 'non-separable and usually non-linear way' interaction is understood in physics (1968, 3).

of languaging to gain the upper hand maintain position or secure cooperation.[36] Capacities that evolved in verbal and gestural exchange with others allow also for Gerd Gigerenzer's 'fast and frugal' responses in human interactions, seemingly without thought, to dangers and opportunities now as once upon a time (2004, 2008). 'Programs' that we inherit or develop can then simply be 'run' without time- and resource-consuming deliberation whenever needed. Our proto-automation of responses (hear the echo of the computational model) renders them habitual, increasing chances of survival, but also creating a problem against which the creative urge rebels. More about that later.

The use of conversation as a benchmark and model for the machine began as early as Turing's 'imitation game' (1950; cf. Gandy 1996). Some years later, Joseph Weizenbaum demonstrated with ELIZA how urgently and readily we attribute true conversational abilities even to a rudimentary simulation.[37] Nevertheless, the realization 'that computational artifacts just are interactive, in roughly the same way as we are' took years more to develop.[38] In the 1990s, CA came into focus as a source for interaction design, but as Lucy Suchman put it, 'remarkably little substantive effect on discourses and practices of the so-called conversational machine' has resulted.[39] In the two collections she cites approvingly, it is notable that attention to the differences between human conversation and digital communication is rare.[40] Ignoring the differences in order simply 'to get on with it' is not futile, but it bypasses questioning the relation.

How, then, does the analogy to conversation help us with the human–machine relation? First and simplest of all, it connects that elusive relation to an extensive body of scholarship elsewhere – and gives back the non-linear possibilities of software as a model for the unobservable cognitive events of talk-in-interaction. Second, it juxtaposes two black-boxed (or semi-opaque) sites from which something new arises, and in doing so aligns and brings both unknowns to the attention of both fields. Third, it illumines the social, competitive, political role that the rule-bound machine (with no small amount of desiring and fearful accommodation) is already filling – and in this role suggests the possibility of an adaptive alteration of human intelligence over time in interaction with the machine as their *relational* intelligence evolves.

Alignment of conversation with digital enquiry is about far more than the design of interfaces. I will have more to say about the alignment of unknowns later.

---

[36]Whiten (1999), Byrne (1996) and Baron-Cohen (1997, 13–20).

[37]Weizenbaum (1976). On current sophistications (e.g. Alexa), see Karppi and Granata (2019) and Levesque (2014).

[38]Suchman (1998, 5). Cf. Norman and Draper (1986), Shneiderman (1980) and Kay (1970); also note 44.

[39]Suchman (2007, 30 n. 2). Since then see Woodruff and Aoki (2010), Perez-Marin and Pascual-Nieto (2011) and Tudini and Liddicoat (2017).

[40]See esp. Button (Chapter 4) and McIlvenny (Chapter 5) in Luff, Gilbert, and Frolich (1990) and Button and Sharrock (Chapter 6) in Thomas (1995). Note Weaver (1949) for the unfortunately persistent confusion of conversation with exchange of 'information'.

5. Cognitive-historical studies of experimentation

Although we may be able to pick out a logical step-by-step 'investigative pathway' in scientific research after the fact, normally it is not known from the beginning but made up, advanced uncertainly, often against unpredictable obstacles and with unanticipated changes of direction (Holmes 2004: xvi). Experimentation is a kind of wayfinding rather than map-following. It is, if you will, a strenuous *conversation* with matter – something Francis Bacon understood, as we will see.

For the historian of experimental science, focusing on a pathway may illumine significant patterns but tends also to direct attention away from the messiness of typically open-ended, experientially driven research, or what geneticist François Jacob has called 'night science' (1998/1997, 126). Keller has argued that from the seventeenth to the nineteenth centuries 'the first-person narrator of the scientific text [was] effectively replaced by the abstract "scientist" ... who could speak for everyman but was no-man'. Thus 'the embodied crafter, interpreter, and reporter of experiments ... the viewing, acting, and doing subject' was effectively erased from the standard account (1996, 418–19) and the cognitive processes involved in the research obscured.

I am concerned to undo the disastrous exclusion of 'night science' – without turning away from 'day science' – in order to throw light on the relation between enquirer and machine. Especially useful for my purposes is the 'cognitive-historical' approach to laboratory work, which proceeds speculatively from 'detailed records of problem-solving behaviour found in diaries and laboratory notebooks' towards the thought-processes responsible for the traces they record.[41] There are, of course, problems. Two of these, the self-reporting bias of the experimenter and dependence on a close fit of historical and contemporary ways of reasoning,[42] are serious enough but I leave them alone here. The third seems to me too highly problematic to pass by: elusiveness of a coherent theory for what happens.[43] Much depends on the meaning of both 'coherent' and 'theory' – surely we do not want a theory that attempts to nail down the creative imagination, rather a way of seeing (*theoría*) as clearly as possible to the horizon. The scholarship from laboratory notebooks in the natural sciences suggests that these problems could gain a new wind once, or if ever, we were to have the equivalent from the computational enquiry of scholars.[44] The hope for progress rests on conceiving this enquiry as exploratory use of technoscientific

---

[41]The term 'cognitive-historical' is Nersessian's (1987); cf. Gooding (2004, 580). Studies with a strong psychological focus on cognition are particularly valuable for my purposes; see e.g. Feist and Gorman (2015, Chapters 4 and 15), Gorman (2009), Gorman et al. (2005), Giere (1992), Gooding (1990) and John-Steiner (1985). I have not made much use of studies orientated primarily to history or to biography.

[42]For self-reporting bias, see Klahr (2000, 8) and Nersessian (1992, 36). For the fit of historical to contemporary reasoning, Nersessian (1992, 36) and Tweney (2013, 87f); also Simon (1986/1983, 159), note Einstein (1950/1936, 59) and Osbeck et al. (2011, 221–6).

[43]Holmes (2004) is particularly concerned. For his attempt at a synthesis of his many historical studies of individual experimenters, see p. 98, also Chapter 1.

[44]My interest here is in the psychology of scholars using the machine to do their research; the psychology of programming and programmers is tangential at best; Hjørland (2013) comes close but still is not it.

equipment to 'experiment' (in the broad sense of that word), though the many differences in assumptions, methods and criteria of the natural vs the human sciences must be taken into account.

For my purposes, the best example for experimental science is David Gooding's landmark studies of Michael Faraday's meticulously kept laboratory notebooks (among the most detailed that have survived).[45] Gooding distinguishes three phases of exploratory enquiry, of which I will consider mostly the first. These are: interactive manipulation of phenomena; their isolation from the instruments and actions which produce them; and the 'semantic ascent', or shift from 'a private world of percepts and objects to ... a public world of talk about objects'.[46] The crucial matter here is the role which first-phase agency, ambiguity and uncertainty play in the science that comes from this work. Hans-Jörg Rheinberger, quoting Jacob's metaphor, describes what goes on during it as combinatorial play: 'an exploratory movement, a game in which one plays with possible positions, an open arrangement ... a game of combinations still unrestricted by the rigorous limits of stringent compatibility' with established findings.[47] The central question is how we get close to the liminal play out of which communicable moves emerge? How do we find out what happens inside the play of the game?

Gooding's keyword for this ontologically ambiguous activity is 'construal':[48] more action than thing, a phenomenon in the active sense of something an experimenter brings to light or makes appear (*phaínei*). Gooding orbited the idea again and again:[49] as a name for a feeling of a way towards a communicable representation of experience during experiments, prior to interpretation, at the pre-hypothetical stage (1990, 71); as a quasi-linguistic messenger 'between the perceptual and the conceptual' (1986, 208), impossible to grasp 'independently of the exploratory behaviour that produces it or the ostensive practices whereby an observer tries to convey it. ... ' (1990, 87); as an action that '*creates* communicable representations of new experience and at the same time integrates these into an existing system of experimental and linguistic practices' (1990, 87). But an unanswered question remains: how the tried and tested construal, accepted and integrated into the science of the day, somehow 'retains enough of its anomalous character [enough of its 'shock of the new'] to promote changes in a ... system that has apparently assimilated it'.[50] To what extent, I wonder,

---

[45]See also esp. Steinle (2016/2005), on Gooding's work, 11, 245–7, 322 and esp Chapter 7; cf Holmes (2004, 96–8). On the notebooks, see Gooding and James (1985).

[46]Gooding (1986, 209), where, as elsewhere, he uses Quine's term (Quine 2013/1960, 249–54).

[47]Rheinberger (2010, 246f), cf. Steinle (2016/2005, 1–3). Such is the way, I would hazard to guess, in which using computers encourages us to talk.

[48]See esp Gooding (1986 *passim*; 1990, xv–xvi, 23, 25–7, 74, 82, 85–8, 115–16, 124–8, 142, 271; 1992a, 102–4), Gorman et al. (2005, Chapter 9), Gooding (2007) and Gooding and Addis (2008). Cf. Tweney (2002, 288, 300), Gorman et al. (2005, Chapter 7) and Beynon (2012). Gooding evidently seems to have taken this term from social psychology of the 1950s; see Ross, Lepper, and Ward (2010, 8).

[49]Cf. Hacking's quite similar 'topical hypotheses' (1992, 45); Tweney's 'inceptions' (2013, 85); Nersessian's 'generic abstractions' (2002, 2008, 191–200).

[50]Gooding (1990, 87 and 29). On 'the shock of the new', see Hughes (1991/1980). Note Gooding's parallel with the arts, 2003: 262, and cf. Gombrich (1961, Chapter 6).

is this anomalous character in all things potentially? Is it, as I have hinted, visible when we are knocked back from knowing a thing to the untrammelled perceiving that precedes knowledge of it?

Gooding's meticulous attempt is ultimately frustrated, hence the orbiting that never lands and the irresistible turn to metaphor. Simply put: he halts at the threshold 'of actions in material and mental space' from which construals emerge and declines to go further (1990, 142). Although he disappoints, Gooding is in good company. The elusiveness of his protean 'construal' marks a threshold beyond which psychology has struggled convincingly to go in its long history of taking up metaphor after metaphor.[51] James Clerk Maxwell likewise turned aside from the 'still more hidden and dimmer region where Thought weds Fact'. 'Does not the way to it pass through the very den of the metaphysician', he asked, 'strewed with the remains of former explorers, and abhorred by every man of science?' (1965/1870, 216). His solution was to use analogy, 'the partial similarity between the laws of one science and those of another which makes each of them illustrate the other' (1864/1855–6, 28). Changing what needs to be changed ('laws' especially), this approaches what I am doing here by aligning digital enquiry analogically with more thoroughly explored kinds.

> 'Emergence' is often used when something intelligible appears from origins into which we have no direct insight. The analogy of experimentation gives us an example. I have so far emphasised the solitary and personal phase, but experimental results must also demonstrate reliability and acceptance; they will inevitably have political implications if not effects. Thorough discussion of these I must leave to another occasion.

6. Divination

Divination is a highly diverse, culturally institutionalized and learned practice, 'extremely widespread, possibly even universal' to *Homo sapiens* from the beginning.[52] In a nutshell, the diviner uses rule-governed apparatus, his or her own body or spontaneous events, to elicit or discern responses to a client's dilemma from forces beyond human control, interpreting these responses with reference to a corpus of traditional knowledge. Here, I am interested specifically in those practices in which the diviner manipulates physical tokens, then draws meaning or an index to it from their configuration.[53]

---

[51]Danziger (2008, esp. Chapter 2), Draaisma (2000/1995, esp. Chapter 6); Leary (1990) and cf. von Neumann (1945, esp. Section 12).

[52]Flad (2008, 403). For work on divination prior to 2005, see Johnston and Struck (2005, 1–10) and Lloyd (1989, 38–49); subsequently, e.g. Flower (2008), Johnston (2008), Flad (2008), Raphals (2013), Holbraad (2012), Zeitlyn (2012), Rochberg (2016) and Struck (2016); several entries in Selin (2016) and Loewe and Blacker (1981). See also Andrew (2018, esp. Chapters 2 and 3).

[53]Plato (*Phaed.* 244cd) and Cicero (*De div.* I.vi.11) divide divination into two kinds: the artificial or technical sort, which relies on human intervention and skill, and the natural, dependent wholly on unintentional, consciously unmotivated phenomena. For challenges to this division, see Rochberg (2016, 24), Zeitlyn (2012, 327), Flower (2008, 84–91), Struck (2016, 16–19), Raphals (2013, 1–2), Johnston (2008, 9) and Peek (1991, 1–22).

Historians and anthropologists have long discredited the view that divination is perilously pseudo-scientific. Nevertheless, this view is apt to gain new life from my alignment of divination with computing, so I had better answer for my use of it here.

Thus, to clear the air: *we do not need to believe in divination* in order to understand how it works, its crucial roles in the societies or communities in which it is practiced and in the historical development of science (Burkert 2005, 36). Belief – the 'mental conviction' that a divine or supernatural authority is manifested in the observables and 'trust, dependence, reliance, confidence, faith' in that authority (*OED* s.v. 1, 2) – is important here only as an historical, cross-cultural fact.

In those terms, then, divination has shown itself capable of systematic, rational and effective enquiry, with much intellectual progress to its credit (Lloyd 2002, 23). It has offered 'established forms of modeling reality and social interaction, of dealing with crisis and conflict … doing so with a high degree of rationality' (Burkert 2005, 30). It has also been bravely adventurous, advancing 'the ambitions of curiosity' (Lloyd 2002) in 'an attempt, perhaps a desperate attempt, to extend the realm of *ratio*, the realm of knowledge and control, beyond the barrier of the future, and the barrier of death, into the misty zones from which normal knowledge and experience is absent' (Burkert 2005, 30). Indeed, the drive to seek answers that human power fails to achieve would appear so strong, persistent and trans-cultural as to suggest that formal aspects of divination can inform, or at least cast light on, our most recent methods of research, including by digital means. What can we infer from divination about the value of enquiry by randomized mechanical means? What does it have to teach us? What has it done for those who have sought its help that other means do for us now?

A number of formidable challenges stand in the way of reliable answers: the tacit assumptions we are likely to project; the reliability of available evidence and the clarity of our objectives.

First, our likely assumptions challenge an attempt to reach beyond historically provincial selves. Unavoidably, or nearly so, we come to divination after a century and a half of having cleared a space for chance (Hacking 1990, 1), worked to tame it and, as a result of that, given to 'random' a key meaning in how we conceive the world. Earlier I focused on its technical meaning in complexity theory and in applications to the physical sciences. But, as I hinted, it has had much broader effects. These are strikingly visible, for example, in the reception history of the Roman poet Lucretius' poetic treatise *De rerum natura* from Marx, Coleridge and Maxwell onward across the natural sciences, arts and humanities,[54] and the board game as a figure of

---

[54]*De rerum natura* 2.289-93 and 254–7; see Fowler (2002, 342–7, 364–6), and on the overall interpretation, 327–30. A brief summary of the Epicurean argument is provided by Lloyd (1973, 23–5); for a fuller treatment, see Long (2006, Chapter 8), and O'Keefe (2009). For the reception history, see e.g. Lezra and Blake (2016), Holmes

thought in the natural sciences and a central model for intelligence in AI. In both, freedom is made possible by unpredictable occurrences or deviations occurring within a tightly ordered system: in Lucretius, by the *clinamen*, or random 'swerve' of atoms that at uncertain places and times breaks the decrees of the Fates and so makes free will possible; in the board game by a randomizing device, such as dice, or by the player's liberty of choice. Thirty years ago, contemporaneous with the fall of the Berlin Wall (1989), the dissolution of the Soviet Union (1991) and public release of the World Wide Web (1991), Hayles conjectured 'that disorder has become a focal point for contemporary theories because it offers the possibility of escaping from what are increasingly perceived as coercive structures of order' (1990, 265). She went on to note that favouring chaos energizes the reactive drive to order; a 'convoluted ambiguity' results, shifting with time. Thus, René Thom's apoplectic reaction ten years earlier, '*Halte au hasard, silence au bruit*' (1980), was not so much a rearguard action as a harbinger. Puzzling out our tendencies of thought from the convoluted ambiguities in the present is the first step in clearing a space for a study of how diviners, their clients and those who have written about them understood what they were doing (Lloyd 1990: 1–2).

Second, the study of divination is challenged by the reliability of evidence from a wide diversity of practices over millennia and across cultures, with an influence beyond reckoning. As Lloyd notes for cross-cultural study of ancient sources, the evidence we have has been filtered through numerous layers of reportage and differences in terminology and interpretation (2002, 1–2).

Third, the objective to see coherence in the variety of phenomena for purposes of comparison is difficult and perilous, to put it mildly. Lloyd has spelled out many if not all of the challenges under three questions that need answering: what is to be compared; what the questions are and what any such study can hope to achieve (1996).

My comparison, as noted earlier, is to the diviner's skilled manipulation of physical tokens in order to read from the resulting configuration a response to a client's dilemma. By focusing in this way, I have already taken my first bridge-building step, shaping the phenomena of divination into a conceptual model suitable to my purposes. My second step is to ask which of the well-attested features of divination come closest to the typical structure of digital enquiry. Approximately in the order I follow these are: (1) the diviner's equipment, method and aim; (2) the client and the question posed; (3) the outcome and (4) the gods or cosmic order to which appeal is made. The resulting model, like any such thing, is necessarily a simplification to be achieved at a minimum level of abstraction from what is being modelled.

---

and Shearin (2012), esp. the Introduction and chapters by Rzepka, Meeker and Holmes; Motte (1986). An online search will yield many examples of twenty-first-century Lucretian invocations.

The questions I am asking sum to this, previously asked but answered only in part: what happens when an enquirer uses 'randomizing' instruments to seek insight about something? What are the processes of thought involved? Why bother? Clearly, great differences separate the terms of the relation between computation and divination, but in both, as earlier, answers are sought deliberately by unpredictable means from sources assumed to be authoritative.

The hope of my stripped-down study is to put as much enlightening stress as possible on the lurking notion of discovery as the uncovering of something already there, ready and waiting. Earlier I suggested that the key to genuinely new and surprising results is the paradox of expectation: not simply the need for a fixed notion so that it may be violated, but the active role of something anticipated in the formation of something unexpected. Here, arguments to the effect that knowledge is in general 'made' or 'crafted' come in, and so Nelson Goodman's key exploration of how a world can be said to be well-made (1984, 30–9). What is the craftsmanship of knowledge-making with digital tools? The process of interest here is rational and systematic, but in the instances of most interest, it is very carefully followed in order to enable something other to happen, something that escapes a rational net yet turns out to be useful.

In the literature, randomizing in divinatory practice is said to accomplish two things:[55] to isolate the outcome of the ritual from prejudicial interference by the participants, providing resistance against a desired outcome, and to allow for communication with those forces beyond human control (a point to which I will return).[56] There are many complications: repeated asking of questions to get the desired answer;[57] the diviner's ambiguous role, sometimes derandomizing by spelling out a meaning, sometimes randomizing further by shifting the meaning (Johnston 2001, 109) and, for the Greeks and Romans, further randomizing by the fickle will of the gods (Raphals 2013, 147). Then there is a reliability of the response, especially problematic for a practice that begins in uncertainty and seeks answers from an imperceptible and perhaps indifferent authority by employing uncontrollable means. In predictive divination reliability is as straightforward as it gets. In ancient Mesopotamia,[58] for example, familiar markers of a science in our sense (or close to it) are evident in the careful observation of celestial phenomena and accumulation of reliable knowledge that make it possible. Reliability becomes much more

---

[55]On divinatory randomization, see Zuesse (1987) and Aubert (1959). In Greco-Roman divination, Flower (2008, 90, 221) (but cf. Raphals 2013, Chapter 5); Johnston and Struck (2005), Introduction (15–16) and chapters by Burkert (37f), Graf (60–2), Grottanelli (134–5), Frankfurter (235) and Johnston (299), Johnston (2003 and 2001: 109–13), Maurizio (1995, 81–6) and Lloyd (1989, 38 n. 120). In Chinese divination, Smith (1991, 19, 204) and Ahern (1981, 53). For anthropological studies of randomization in contemporary cultures, see Holbraad (2012, 149–50), Dove (1999, 378 and 1993, 146–7, 151–2), Zeitlyn (1995), Peek (1991, 203–4), Dove (1983), Park (1963, 198–200), Moore (1957, 71–3) and cf. Graw (2009).
[56]Johnston (2005, 299), cf. Johnston and Struck (2005, 15), Maurizio (1995, 81), Park (1963: 198f) and Dove (1993).
[57]Graf (2005, 52) and Aune (2004, 371). Cf. Kim (2018, 373).
[58]Raphals (2013, 2–3), Lloyd (2002, Chapter 2).

problematic when that which is to be divined involves biological, psychological or social phenomena.

We have to be cautious in seeing our knowledge practices in those of others. I just cited the close resemblance of modern science to predictive Mesopotamian divination, but it was practiced not to gain knowledge for its own sake, rather to foresee the future and so advise the monarch (Lloyd 2002, 28). We are apt to think that the ancients undertook individual consultations as we would, out of 'purely private interests in a modern, Western sense of the term', but this was rarely if ever the case for the Greeks and Chinese, for example.[59] Their motivating desire may have been to get out of a jam, or to be prepared for a difficulty, or to reach for the universal *sympatheia* that Walter Burkert sees as an ultimate objective. But they were not ruled by the same 'egocentric system' of modern lifestyle choices and agonies (Burkert 2005, 48; cf. Giddens 1991). Referring to the injunction to self-knowledge at Delphi, Peter Struck comments that 'for the Greeks it was not so much a question of knowing oneself as a sui generis individual but rather of developing an understanding of oneself as a member of the order of things'.[60] Alignment with and on behalf of others, alignment to the cosmic order, or the will of a god or the gods, however variable, was the norm.

Divination that was or is sought for the wisdom to attune oneself to the order of things brings us closer to the ends of exploratory enquiry in the human sciences and so is of particular interest here. Though doubtless sometimes confirmatory, attunement to a transcendent order is a serious, quite possibly upsetting matter, *mutatis mutandis* like the twentieth-century listener's assimilation to 'the shock of the new' delivered by 'the advent of musical modernism', for which the violent surprise caused by Stravinsky's *Rite of Spring* in 1913 is an example: a briefly estranging, traumatic, self-changing encounter.[61]

7. Alignment and relation

Now, as noted earlier, I want to look again at my three analogies to ask what kind of relation we might conceive by *aligning* conversation, experimentation and divination to digital enquiry. What were or are the practitioners of these activities doing or trying to do that illumines our uses of the machine? What about all of these casts light on the question of how we conceive and exercise intelligence, biological and artificial, in its use?

---

[59]Raphals (2013, 251); Rochberg comments for the Assyrian material that, 'There's virtually no evidence of private divining, and even when personal horoscopes come into the picture it's entirely unknown how far beyond the very most elite people the whole thing went' (private e-mail, March 25, 2019). Cf. Starr (1990). For the overall point, see Lloyd (2002, 21) and Vernant (1991/1974, 303).

[60]Struck (2016, 2). Cf. Strathern's use of the old term 'dividual' (OED 3, 'distributed among a number ... held in common') in the context of recombinant Melanesian personhood (1999, 60) and Keller's argument for 'an innate (and near-universal) capacity for self-reflection underlying experiences of first personhood ... of a core subjectivity', the turbulent changes afoot in how we construct ourselves and our ability to press this radically (2007, 354).

[61]Note esp. Huron's analysis of randomization in the divinatory rite (2007, 344–6); cf. Taruskin (2003, 281–3).

7.1. Conversation

Sociolinguistic analysis of ordinary conversation has shown it to be complex and unpredictable. Studies in the origins of hominid intelligence suggest its crucial evolutionary significance in the development social adaptation and negotiation. Conversation as metaphor and model supplies a nearly ubiquitous way of probing the relation of human to artificial intelligence and challenging efforts to implement it. But the evidence for how talk-in-interaction does what it does in meetings of minds seems out of reach (as it is in the analogous situations of digital enquiry, laboratory science and divination). So, we are forcibly returned to the question Erving Goffman asked of puzzling social encounters: 'What is it that's going on here?' (1974, 8)

We must be careful not to allow the attractive Machiavellian Hypothesis or in general the emphasis on doing things with words to obscure the kind of talking Gadamer called a 'genuine conversation … [which is] never the one that we wanted to conduct', rather the kind we fall into, the kind we are led by rather than lead, the kind that surprises rather than fulfils expectations or replays a conventional script (Gadamer 2004/1989, 385). *This* kind points to a metaphorical space between the two differently constituted agents created by their rapid back-and-forth exchange in which the intelligence of each is augmented by the affordance of the other. The analogy of conversation, that is, aligns to thinking which happens beyond the brain, in and with the world, and to computing that likewise takes place beyond the interface, in the niche augmented by the machine-in-use.[62]

On the face of it, even in the simplest exchange, face-to-face talk sets a *very* high bar. Computing systems can mimic verbal conversation and will undoubtedly get better at it. We may want to draw a line, but the attempt to fix one absolutely between human and artificial performance is not just in vain, it misses the point: *both are in reciprocal, co-evolutionary development.* Much more productive is to emphasize the figurative sense of conversation as 'Occupation or engagement *with* things' (*OED* 4), taking up the invitation to stretch the term beyond the observable to what we do in active and intimate relations with the people, institutions and material things that matter to us. Consider conversational relations in Tim Ingold's anthropology, in the human geography of landscape and in the work of early cybernetic artists and theorists,[63] whose projects strove to implement as well as understand the work of art in Gadamer's later view: 'a fruitful conversation, a question and answer … , a true dialogue whereby something has emerged and "remains"' (1985, 250). Intelligence *arte factum* – a *künstliche Intelligenz* – may not be here yet, but its non-imitative

---

[62]For the brain, Clark (2008) and Anderson (2014). For the machine, Goldin and Wegner (2006); note the latter's use of 'autistic' in Wegner (1998).

[63]Ingold (2011), Benediktsson and Lund (2010), Abram (1997), Pask (1971), Reichardt (1968) and Brown et al. (2008).

conversation must already be happening and, as I've argued, needs our attention to develop along its native lines.

'In inner speech and in conversation', Goody observes, 'dialogue and the dyad are built into human cognition' (1995, 12; cf. Simmel 2009/1908, 82–114). We talk to ourselves, to each other and sometimes, finding chance or other imponderables difficult to accept, posit a conversational Other (Goody 1995, 207–8). Or, we design and build them. The question now is how to design the things that we build to initiate conversations that would force us to rethink thinking – and rethink science – once again.

7.2. Experimentation

The parallel between digital enquiry and laboratory experimentation is close. But consider where this leads us: on the one hand, to examples of scholarship and artistic work exploring the relation of cultural artefacts to algorithmic 'natural law';[64] and on the other to the pluralization of 'ontology' in philosophy and computer science and, following Galison, the progressive substitution of the computer for 'nature itself' (1996, 157; McCarty 2018). But we can go even further. Lloyd's historical argument that 'nature' is itself an invention shows 'natural law' to be doubly metaphorical: codified judicial decision-making transferred from the court to the cosmos, historically in some instances by way of a divine judge. Perhaps we can say, then, that the digital machine has prompted or furthered the rediscovery of nature as a contingent hypothesis, or as Bacon wrote in *De sapientia veterum* (1617), a Proteus with whom we are in perpetual struggle.

We've seen that digital enquiry, analysis of conversation and laboratory experimentation reach towards the unknowable but partially controllable source of new insights. And so Goody calls for more work on the common ground between interacting talkers; Gooding declined to enter the 'material and mental space' from which insights come; Maxwell wrote of a metaphysician's den, from which he advised we turn away. But that's not the end of it. There is (as anticipated) one further step to take along this path.

7.3. Divination

Having isolated the uncomputable by devising the eponymous abstract mathematical device on which the digital machine was later based (1936/7), Turing then imagined a special machine incorporating an 'oracle' to enable it to solve problems as a human mathematician would.[65] He proposed this

---

[64]For the creative arts, in addition to footnote 63: Funkhouser (2012), Boden (2011), Argamon, Burns, and Dubnov (2010), for literary studies, Burrows (2010), in which the 'balance between algorithmic analysis and questions of literary importance' is exemplary (David Hoover, private e-mail, November 27, 2019); Hoover (2016), Moretti (2013) and the pamphlets of the Stanford Literary Lab (https://litlab.stanford.edu/).
[65]Turing (2012/1938, 52–3) for his mention of the 'oracle'; note there the brief comments by Appel (7) and Feferman (22–3).

oracle not as an analogue to any human mental faculty (for, in Andrew Hodges' words, 'it does something no human being could do') but to facilitate the study of 'the mental "intuition" of truths which are not established by following mechanical processes'.[66] Later, four developments came to approximate or model the oracle's function: the non-linear design of digital hardware; the introduction of randomizing software (cf. Mirowski 2002, 148); online computing, in which the machine interacts with and is affected by the world (Soare 2009, 387–9); and, in game-playing and 'deep learning', the pattern-matching and self-modification which can come up with something that, as Turing foresaw, surprises us. In my sense, these latter two move the locus of machine 'intelligence' closer towards that cognitive, physical and social space where intelligence becomes relational (footnote 2). Conversation (again, setting its *very* high bar) suggests the further challenge posed by mistakes and their repair. Error that makes sense in context may, in the end, prove the greatest hurdle.

In the anthropology of divination, 'random' likewise steps in to gloss the outcome of the diviner's actions, which in Graf's words 'introduces a gap where the hand and mind of the divinity can interfere' (2005, 60–1). I prefer to avoid the common rationalization while respecting the sources, saying not that a divinity reaches down and *interferes* in the mechanics of casting tokens but that divine will or cosmic order is *manifested in, revealed by* and *read from* that which we call the 'randomness' of a configuration – and that (expecting the unexpected) first the diviner, then the client *makes* of or from the result what he or she can.

Looking now to computing from that perspective, it should be obvious that we do not need a corresponding ghost in the machine. As complexity theory tells us, the machine in its applications to ever more complex problems yields an ever clearer glimpse of a physical world not to be nailed down by an eventually complete body of 'natural law'; however useful, and productive this legislative metaphor may have been and may continue to be. Curiosity's ambitions exceed it. The machine that we have gives us an oracle by which a tamed chance can be put powerfully to work.

8. In illo tempore

In the three analogical cases I have reviewed, and in digital enquiry, an outcome undergoes 'the test of shareable experience', first by the participants, who act in the knowledge of their own constitution as persons in the regard of others, then by those other 'political animals', to which each is answerable.[67] But if the result is proved reliable and important, what then happens?[68]

---

[66]Hodges (2013, 15–17); cf. Cooper and Hodges (2016: Chapters 13–15).
[67]Gooding (1990, 85), Strathern (1988, 275) (paraphrased) and Aristotle, *Pol.* 1253a, respectively; cf. Fleck (1981/1935) and Keller (2007, 354).
[68]Latour and Woolgar (1986/1979), Gooding (1990; 1986). Cf. Goldman and O'Connor (2019); the journals *Social Studies of Science* and *Social Epistemology*.

Consider the epigrapher in Graf's account, who halts before an oracular pronouncement, 'rather puzzled by the occurrence of what we only can understand as scribal errors' but aware that the ancient reader was able to make sense of it (2005, 78). Graf celebrates 'the human capacity to find meaning in what seem to be or really are random phenomena'. But (as twice anticipated so far), there are two problems here, marked by 'find' and 'really are', both referring back to the ambiguity of 'invent' (*invenire,* to come upon, to make) that has been central to this paper throughout. Did the ancients *find* this new meaning, or *make* it, or did it *occur to* or *in* them, stimulated by those inscriptional signs the grammarian finds erroneous, and they meaningful? Again, what is it that's going on here?

In the cognitive sciences from Donald Campbell's 'blind variation and selective retention' (1960) to recent arguments for a 'predictive brain',[69] debate about this continues to circle the question of whether the new comes about through a 'sighted, guided, or directed' effort or a blind one (Simonton 2011, 159). In intercourse with the machine, it is surely both: the product of a blind-but-designed combinatorial cogitation triggering the curious mind – to 'invent'. But what grabs us about the invented, exactly?

I have called its relevant quality 'new', but – I ask this again – how does anything 'new' qualify as such? Responding to this question in the workshop at which an earlier version of this essay was presented, Lloyd energetically asked in turn, 'What are the conditions of identity of a thought?'[70] What makes a new one new, considering that the sensorium is abuzz constantly with novel perceptions, the healthy mind filtering them to avoid hyperaesthesia but also staying alert for opportunities to improve the filtering, to let better ones (or better opportunities to fashion good ones) in? Gooding, you will recall, has asked and marked as unanswered the question of how the anomalous character of a new result survives assimilation so as to retain a yeast-like power to effect change. But perhaps we have the question the wrong way around. And I have asked, might this be true of all things?

It is time to stop swerving past the answer at which I have been aiming all along. My title, from the Russian Formalist Viktor Shklovsky's 'Art as technique' (1917), betrays it. In that essay, Shklovsky shakes his fist at comfortable, predictable but deadly habitualization – the algorithmization of life, we might say. He offers a way out – momentary because we perpetually fall back into habit:

> art exists that one may recover the sensation of life; it exists to make one feel things, to make the stone stony. The purpose of art is to impart the sensation of things as they are perceived and not as they are known. The technique of art is to make objects 'unfamiliar,' to make forms difficult, to increase the difficulty and length of perception

---

[69]On the career of Campbell's BVSR hypothesis: Simonton (2011), the debate in *Physics of Life Reviews* 7.2 (2010) with Simonton's reply referring to 'combinatorial models of exceptional creativity' (190–4); on the predictive brain: Yon, de Lange, and Press (2018); cf. Anderson et al. (2016) and Clark (2016) (in the present context note esp. p. 79 on 'very (agent-) surprising things', p. 129 n. 15 and Section III); Clark (2013).
[70]'Science in the Forest, Science in the Past II', Needham Research Institute, Cambridge, 14 June 2019. On the new, see North (2013), D'Angour (2011), March (2010), Crosby (2009) and Strathern (2005).

because the process of perception is an aesthetic end in itself and must be prolonged. (1965/1917, 12)

'Aesthetic ends' may not be what a digital enquirer would think of, but Shklovsky's остранение (*ostranenyi*), the 'defamiliarisation' of the familiar, rendering the ordinary uncanny, 'rendering the already nameable *only just* un-nameable'[71], is what the work of art can do, and what the *künstliche Intelligenz* I have been discussing can accomplish in the space of conversation with us (cf. McCarty 2019, 154).

Shklovsky wrote his timely but very old call to action at the beginning of the last century. Fifteen years ago, Burkert recommended the oracular sign's power to help the ancients 'to get out of a closed, egocentric system, to get into touch with "otherness," with the whole environment, to experience the all-embracing net of existence, nay universal *sympatheia*', then commented: 'This ought to challenge even the noisy self-resonance of contemporary society'. At the beginning of 2020, contemplating a telling expression of the then current political scene, James Lasdun urged on us as antidote 'anything that can shed light … on the processes by which groups of people seal themselves into airtight alternative realities' (2020, 27). I have argued that the relational intelligence of the human–machine coupling puts into our hands a tool of unsealing power at a time when its influence could make a difference. Our move.

Disclosure statement

No potential conflict of interest was reported by the author(s).

References

Abram, David. 1997. *The Spell of the Sensuous: Perception and Language in a More-Than-Human World*. New York: Vintage Books.
Ahern, Emily Martin. 1981. *Chinese Ritual and Politics*. Cambridge: Cambridge University Press.
Alligood, Kathleen T., Tim D. Sauer, and James A. Yorke. 1996. *Chaos: An Introduction to Dynamical Systems*. New York: Springer-Verlag.
Anderson, Michael L. 2014. *After Phrenology: Neural Reuse and the Interactive Brain*. Cambridge, MA: MIT Press.

[71]Tim Smithers, June 29, 2020.

Anderson, Michael L. et al. 2016. "'Précis of *After Phrenology: Neural Reuse and the Interactive Brain*' with Open Peer Commentary." *Behavioral and Brain Sciences* 39 (16 June): 1–45.

Andrew, Christopher. 2018. *The Secret World: A History of Intelligence*. New Haven, CT: Yale University Press.

Argamon, Shlomo, Kevin Burns, and Shlomo Dubnov, eds. 2010. *The Structure of Style: Algorithmic Approaches to Understanding Manner and Meaning*. Berlin: Springer-Verlag.

Ascher, Marcia. 1991. *Ethnomathematics: A Multicultural View of Mathematical Ideas*. Pacific Grove, CA: Brooks/Cole.

Aspray, William. 1990. *John von Neumann and the Origins of Modern Computing*. Cambridge, MA: MIT Press.

Aubert, Vilhelm. 1959. "Chance in Social Affairs." *Inquiry* 2 (1–4): 1–24.

Auerbach, Erich. 2003/1953. *Mimesis: The Representation of Reality in Western Literature. Translated Willard R. Trask*. Princeton NJ: Princeton University Press.

Aune, David E. 2004. "Divination and Prophecy." In *Religions of the Ancient World: A Guide*, edited by Sarah Iles Johnston, 370–391. Cambridge, MA: Harvard University Press.

Austin, J. L. 1962. *How to Do Things with Words*. The William James Lectures delivered at Harvard University in 1955. Oxford: Clarendon Press.

Bacon, Francis. 1617. *De sapientia veterum*. London: Johannes Billius. Accessed 4/11/20 from the Internet Archive, https://archive.org/details/bub_gb_K-1GHYHbnBsC/mode/2up.

Baron-Cohen, Simon. 1997. *Mindblindness: An Essay on Autism and Theory of Mind*. Cambridge, MA: MIT Press.

Beltrami, Edward. 1999. *What Is Random? Chance and Order in Mathematics and Life*. New York: Springer Science+Business Media.

Benediktsson, Karl, and Katrín Anna Lund. 2010. *Conversations with Landscape*. London: Ashgate.

Bennett, Deborah J. 1998. *Randomness*. Cambridge, MA: Harvard University Press.

Berge, C. 1971/1968. *Principles of Combinatorics*. New York: Academic Press.

Berliner, Paul F. 1994. *Thinking in Jazz: The Infinite Art of Improvisation*. Chicago: University of Chicago Press.

Beynon, Meurig. 2012. "Modelling with Experience: Construal and Construction for Software." In Bissell and Dillon 2012: 197–228.

Biggs, Norman L., E. Kenneth Lloyd, and Robin J. Wilson. 1995. "The History of Combinatorics". In Graham, Grötschel and Lovász 1995: 2163–98.

Bishop, Robert. 2017. "Chaos". *Stanford Encyclopedia of Philosophy*, edited by Edward N. Zalta. https://plato.stanford.edu/archives/spr2017/entries/chaos/ (17 June 2020).

Boden, Margaret. 2011. *Creativity and Art: Three Roads to Surprise*. Oxford: Oxford University Press.

Bréard, Andrea. 2015. "What Diagrams Argue in Late Imperial Chinese Combinatorial Texts." *Early Science and Medicine* 20: 241–264.

Brin, Sergey, and Lawrence Page. 1998. "The Anatomy of a Large-Scale Hypertextual Web Search Engine." *Computer Networks and ISDN Systems* 30: 107–117.

Brown, Paul, Charlie Gere, Nicholas Lambert, and Catherine Mason, eds. 2008. *White Heat Cold Logic: British Computer Art 1960–1980*. Cambridge, MA: MIT Press.

Brown, G. Spencer, and G. B. Keene. 1957. "Symposium: Randomness." *Proceedings of the Aristotelian Society, Supplementary Volumes* 31: 145–160.

Burkert, Walter. 2005. "Signs, Commands, and Knowledge: Ancient Divination between Enigma and Epiphany". In Johnston 2005: 29–49.

Burrows, John. 2010. "Never Say Always Again: Reflections on the Numbers Game." In *Text and Genre in Reconstruction: Effects of Digitalization on Ideas, Behaviours, Products and Institutions*, edited by Willard McCarty, 13–36. Cambridge: Open Book Publishers.

Byrne, Richard W. 1996. "Machiavellian Intelligence." *Evolutionary Anthropology* 5 (5): 172–180.

Campbell, Donald T. 1960. "Blind Variation and Selective Retention in Creative Thought as in Other Knowledge Processes." *Psychological Review* 67 (6): 380–400.

Campbell-Kelly, Martin. 2011. "From Theory to Practice: The Invention of Programming, 1947–51." In *Dependable and Historic Computing: Essays Dedicated to Brian Randell on the Occasion of His 75th Birthday*, edited by Cliff B. Jones and John L. Lloyd, 23–37. Heidelberg: Springer-Verlag.

Castoriadis, Cornelius. 1987/1975. *The Imaginary Institution of Society. Translated Kathleen Blamey*. Cambridge: Polity Press.

Chesher, Chris. 1997. "The Ontology of Digital Domains." In *Virtual Politics: Identity and Community in Cyberspace*, edited by David Holmes, 79–92. London: Sage.

Chun, Wendy Hui Kyong. 2011. *Programmed Visions: Software and Memory*. Cambridge, MA: MIT Press.

Ciula, Arianna, Øyvind Eide, Cristina Marras and Patrick Sahle, eds. 2018. "Models and Modelling between Digital Humanities & Humanities – A Multidisciplinary Perspective." HSR Supplement 31. Historical Social Research/Historische Sozialforschung.

Clark, Andy. 2008. *Supersizing the Mind: Embodiment, Action and Cognitive Extension*. Oxford: Oxford University Press.

Clark, Andy. 2013. "Whatever Next? Predictive Brains, Situated Agents, and the Future of Cognitive Science." *Behavioral and Brain Sciences* 36: 181–253.

Clark, Andy. 2016. *Surfing Uncertainty: Prediction, Action, and the Embodied Mind*. Oxford: Oxford University Press.

Clemente, Ignasi. 2013. "Conversation Analysis and Anthropology." In Sidnell and Stivers 2013: 688–700.

Colburn, Timothy, and Gary Shute. 2007. "Abstraction in Computer Science." *Minds & Machines* 17: 169–184.

Conway, J. H. 2001. *On Numbers and Games*. London: Academic Press.

Cook, Roy T., and Sondra Bacharach. 2017. *Lego and Philosophy: Constructing Reality Brick by Brick*. Chichester: Wiley Blackwell.

Cooper, S. Barry, and Andrew Hodges, eds. 2016. *The Once and Future Turing*. Cambridge: Cambridge University Press.

Crosby, Donald A. 2009. "Causality, Time, and Creativity: The Essential Role of Novelty." *The Pluralist* 4 (3): 46–59.

Dale, Rick, Riccardo Fusaroli, Nicholas D. Duran, and Daniel C. Richardson. 2014. "The Self-organization of Human Interaction." *Psychology of Learning and Motivation* 59: 43–96.

D'Angour, Armand. 2011. *The Greeks and the New: Novelty in Ancient Greek Imagination and Experience*. Cambridge: Cambridge University Press.

Danziger, Kurt. 2008. *Marking the Mind: A History of Memory*. Cambridge: Cambridge University Press.

David, F. N. 1962. *Games, Gods and Gambling. The Origins and History of Probability and Statistical Ideas from the Earliest Times to the Newtonian era*. New York: Hafner.

Dawson, Michael R. W., Brian Dupuis, and Michael Wilson. 2010. *From Bricks to Brains: The Embodied Cognitive Science of LEGO Robots*. Edmonton: Athabasca University Press.

Dennett, Daniel C. 1995. *Darwin's Dangerous Idea: Evolution and the Meanings of Life.* London: Penguin.

de Rosnay, Marc et al. 2014. "Talking Theory of Mind Talk: Young School-aged Children's Everyday Conversation and Understanding of Mind and Emotion." *Journal of Child Language* 41: 1179–1193.

DeTemple, Duane, and William Webb. 2014. *Combinatorial Reasoning: An Introduction to the Art of Counting.* Hoboken, NJ: John Wiley & Sons.

Dijkstra, E. W. 1986. "On a Cultural Gap." *The Mathematical Intelligencer* 8 (1): 48–52.

Dove, Michael R. 1983. "Forest Preference in Swidden Agriculture." *Tropical Ecology* 24 (1): 122–142.

Dove, Michael R. 1993. "Uncertainty, Humility, and Adaptation in the Tropical Forest: The Agricultural Augury of the Kantu." *Ethnology* 32 (2): 145–167.

Dove, Michael R. 1999. "Forest Augury in Borneo: Indigenous Environmental Knowledge – About the Limits to Knowledge of the Environment." In *Cultural and Spiritual Values of Biodiversity*, edited by Darrell Addison Posey, 376–380. London: Immediate Technology Publications, United Nations Environmental Programme.

Draaisma, Douwe. 2000/1995. *Metaphors of Memory: A History of Ideas About the Mind.* Trans. Paul Vincent. Cambridge: Cambridge University Press.

Drew, Paul. 1995. In Goody 1995: 111–38.

Duranti, Alessandro. 1997. *Linguistic Anthropology.* Cambridge: Cambridge University Press.

Duranti, Alessandro. 2015. *The Anthropology of Intentions: Language in a World of Others.* Cambridge: Cambridge University Press.

Duranti, Alessandro, and Kenny Burrell. 2004. "Jazz Improvisation: A Search for Hidden Harmonies and a Unique Self"/"L'improvvisazione jazz: Alla ricercar di armonie nasciste e di un se'distinto." *Ricerche di Psicologia* 27 (3): 71–101.

Duranti, Alessandro, Elinor Ochs, and Bambi B. Schieffelin, eds. 2012. *The Handbook of Language Socialization.* Chichester, West Sussex: Wiley-Blackwell.

Edwards, Derek, and David Middleton. 1987. "Conversation and Remembering: Bartlett Revisited." *Applied Cognitive Psychology* 1: 77–92.

Einstein, Albert. 1950/1936. *Out of My Later Years.* New York: Philosophical Library.

Evens, Aden. 2015. *Logic of the Digital.* London: Bloomsbury.

Fairbairn, John. 2007. "Go in China". In Finkel 2007: 133–7.

Faulkner, Robert R., and Howard S. Becker. 2009. *"Do you Know … ?" The Jazz Repertoire in Action.* Chicago: University of Chicago Press.

Feist, Gregory J., and Michael E. Gorman, eds. *Handbook of the Psychology of Science.* New York: Springer.

Finkel, I. L. 2007. *Ancient Board Games in Perspective. Papers from the 1990 British Museum Colloquium, with Additional Contributions.* London: The British Museum Press.

Flad, Rowan K. 2008. "Divination and Power: A Multiregional View of the Development of Oracle Bone Divination in Early China." *Current Anthropology* 49 (3): 403–437.

Fleck, Ludwik. 1981/1935. *Genesis and Development of a Scientific Fact*, edited by Thaddeus J. Trenn and Robert K. Merton. Translated Fred Bradley and Thaddeus J. Trenn. Chicago, IL: University of Chicago Press.

Flower, Michael Attyah. 2008. *The Seer in Ancient Greece.* Berkeley, CA: University of California Press.

Fodor, Jerry A. 1983. *The Modularity of Mind.* Cambridge, MA: MIT Press.

Fowler, Don. 2002. *Lucretius on Atomic Motion. A Commentary on De Rerum Natura Book Two, Lines 1-332.* Oxford: Oxford University Press.

Fox, Barbara A., Sandra A. Thompson, Cecilia E. Ford, and Elizabeth Couper-Kuhlen. 2013. "Conversation Analysis and Linguistics." In Sidnell and Stivers, 726–740.

Funkhouser, Christopher. 2012. "First-Generation Poetry Generators: Establishing Foundations in Form." In *Mainframe Experimentalism: Early Computing and the Foundations of the Digital Arts*, edited by Hannah B. Higgins, and Douglas Kahn, 243–265. Berkeley: University of California Press.

Gadamer, Hans-Georg. 1985. "Philosophy and Literature". Translated Anthony J. Steinbock. *Man and World* 18: 241–259.

Gadamer, Hans-Georg. 2004/1989. *Truth and Method*. 2nd ed. Translated Joel Weinsheimer and Donald C. Marshall. London: Continuum.

Galison, Peter. 1996. "Computer Simulations and the Trading Zone." In *The Disunity of Science: Boundaries, Contexts, and Power*, edited by Peter Galison and David J. Stump, 118–157. Stanford, CA: Stanford University Press.

Gallagher, Richard, Tim Appenzeller, and Dennis Normile. 1999. "Beyond Reductionism." *Science NS* 284 (5411 (2 April)): 79–109.

Gandy, Robin. 1996. "Human Versus Mechanical Intelligence." In *Machines and Thought: The Legacy of Alan Turing. Volume I*, edited by P. J. R. Millican and A. Clark, 125–136. Oxford: Clarendon Press.

Gell, Alfred. 1998. *Art and Agency: An Anthropological Theory*. Oxford: Clarendon Press.

Gibbs, Raymond W., Jr. 2005. *Embodiment and Cognitive Science*. Cambridge: Cambridge University Press.

Gibson, James J. 2015/1979. *The Ecological Approach to Visual Perception*. Classic edn. New York: Psychology Press.

Giddens, Anthony. 1991. *Modernity and Self-Identity: Self and Society in Late Modern Age*. London: Polity.

Giere, Ronald N., ed. 1992. *Cognitive Models of Science*. Minnesota Studies in the Philosophy of Science, Vol. XV. Minneapolis MN: University of Minnesota Press.

Giere, Richard N. 2002. "Models as Parts of Distributed Cognitive Systems." In *Model-based Reasoning: Science, Technology, Values*, edited by Lorenzo Magnani and Nancy J. Nersessian, 227–241. New York: Springer Science+Business Media.

Gigerenzer, Gerd. 2004. "Fast and Frugal Heuristics: The Tools of Bounded Rationality." In *Blackwell Handbook of Judgement and Decision Making*, edited by Derek J. Koehler and Nigel Harvey, 62–88. Oxford: Blackwell.

Gigerenzer, Gerd. 2008. *Gut Feelings: The Intelligence of the Unconscious*. New York: Viking.

Goffey, Andrew. 2008. "Intelligence." In *Software Studies: A Lexicon*, edited by Matthew Fuller, 132–142. Cambridge MA: MIT Press.

Goffman, Erving. 1974. *Frame Analysis: An Essay on the Organization of Experience*. Boston: Northeastern University Press.

Goldin, Dina, and Peter Wegner. 2006. "Principles of Interactive Computation." In *Interactive Computation: The New Paradigm*, edited by Dina Goldin, Scott A. Smolka, and Peter Wegner, 25–37. Berlin: Springer.

Goldman, Alvin, and Cailin O'Connor. 2019. "Social Epistemology". *Stanford Encyclopedia of Philosophy*, Fall 2019 edn. Stanford CA: Center for the Study of Language and Information, Stanford University. https://plato.stanford.edu/archives/fall2019/entries/epistemology-social/ (28 November 2019).

Goldstine, Herman H. 1972. *The Computer from Pascal to von Neumann*. Princeton: Princeton University Press.

Goldstine, Herman H., and John von Neumann. 1947. *Planning and Coding of Problems for an Electronic Computing Instrument. Report on the Mathematical and Logical aspects of an Electronic Computing Instrument*, Part II, Volumes 1–3. Princeton, NJ: Institute for

Advanced Study. https://ia800301.us.archive.org/13/items/planningcodingof0103inst/planningcodingof0103inst.pdf (10 April 2019).

Gombrich, E. H. 1961. *Art and Illusion: A Study in the Psychology of Pictorial Representation*. London: Phaidon.

Gooding, David. 1986. "How Do Scientists Reach Agreement about Novel Observations?" *Studies in the History and Philosophy of Science* 17 (2): 205–230.

Gooding, David. 1990. *Experiment and the Making of Meaning: Human Agency in Scientific Observation and Experiment*. Dordrecht: Kluwer Academic.

Gooding, David. 2003. "Varying the Cognitive Span: Experimentation, Visualization, and Computation." In *The Philosophy of Scientific Experimentation*, edited by Hans Radder, 255–301. Pittsburgh, PA: University of Pittsburgh Press.

Gooding, David. 2004. "Cognition, Construction and Culture: Visual Theories in the Sciences." *Journal of Cognition and Culture* 4: 551–593.

Gooding, David. 2007. "Alchemy, the Calculus and Electromagnetism: Some Historical Encouragement for TTC". *Thinking through Computing*, Department of Computer Science, University of Warwick, November 2007. https://go.warwick.ac.uk/em/thinkcomp07/gooding2.pdf (11 April 2019).

Gooding, David, and T. R. Addis. 2008. "Modelling Experiments as Mediating Models." *Foundations of Science* 13: 17–35.

Gooding, David, and Frank A. J. L. James, eds. 1985. *Faraday Rediscovered: Essays on the Life and Work of Michael Faraday, 1791-1867*. London: Macmillan.

Goodman, Nelson. 1984. *Of Mind and Other Matters*. Cambridge, MA: Harvard University Press.

Goody, Esther N. 1995. *Social Intelligence and Interaction: Expressions and Implications of the Social Boas in Human Intelligence*. Cambridge: Cambridge University Press.

Gorman, Michael E. 2009. "Introduction to Cognition in Science and Technology." *Topics in Cognitive Science* 1: 675–685.

Gorman, Michael E., Ryan D. Tweney, David C. Gooding, and Alexandra P. Kincannon. 2005. *Scientific and Technological Thinking*. Mahwah, NJ: Lawrence Erlbaum.

Graf, Fritz. 2005. "Rolling the Dice for an Answer." In Johnston and Struck 2005: 51–97.

Gramelsberger, Gabriele. 2011. "From Computation with Experiments to Experiments with Computers." In *From Science to Computational Sciences: Studies in the History of Computing and Its Influence on Today's Sciences*, edited by Gabriele Gamelsberger, 131–142. Zürich: Diaphanes.

Graw, Knut. 2009. "Beyond Expertise: Reflections on Specialist Agency and the Autonomy of the Divinatory Ritual Process." *Africa: Journal of the International African Institute* 79 (1): 92–109.

Hacking, Ian. 1990. *The Taming of Chance*. Cambridge: Cambridge University Press.

Hacking, Ian. 1992. "The Self-Vindication of the Laboratory Sciences." In *Science as Practice and Culture*, edited by Andrew Pickering, 29–64. Chicago, IL: University of Chicago Press.

Hagen, Wolfgang. 2006/1997. "The Style of Sources: Remarks on the Theory and History of Programming Languages." Trans. Peter Krapp. In *New Media, Old Media: A History and Theory Reader*, edited by Wendy Hui Kong Chun and Thomas Keenan, 157–175. New York: Routledge.

Hao, Bai-lin. 1989. *Elementary Symbolic Dynamics and Chaos in Dissipative Systems*. Singapore: World Scientific.

Hasslacher, Brosl. 1995. "Beyond the Turing Machine". In *The Universal Turing Machine: A Half-Century Survey*. 2nd ed. Wien: Springer-Verlag.

Hawkins, David. 1946. *Inception Until August 1945. Vol. 1 of Manhattan District History, Project Y, The Los Alamos Project*. Los Alamos, NM: Los Alamos Scientific Laboratory. See https://www.osti.gov/opennet/manhattan-project-history/Resources/library.htm (11 April 2019).

Hayles, N. Katherine. 1990. *Chaos Bound: Orderly Disorder in Contemporary Literature and Science*. Ithaca, NY: Cornell University Press.

Hennessy, John L., and David A. Patterson. 2012. *Computer Architecture: A Quantitative Approach*. 5th ed. Amsterdam: Elsevier.

Hjørland, Birger. 2013. "User-based and Cognitive Approaches to Knowledge Organization: A Theoretical Analysis of the Research Literature." *Knowledge Organization* 40 (1): 11–27.

Hodges, Andrew. 2013. "Alan Turing". *Stanford Encyclopedia of Philosophy*, Winter 2013 Edition. Stanford CA: Center for the Study of Language and Information, Stanford University. http://plato.stanford.edu/archives/win2013/entries/turing/ (28 November 2019).

Holbraad, Martin. 2012. *Truth in Motion: The Recursive Anthropology of Cuban Divination*. Chicago: University of Chicago Press.

Holland, John H. 1998. *Emergence: From Chaos to Order*. Reading, MA: Addison-Wesley.

Holland, John H. 2014. *Complexity. A Very Short Introduction*. Oxford: Oxford University Press.

Holmes, Brooke, and W. H. Shearin. 2012. *Dynamic Reading: Studies in the Reception of Epicureanism*. Oxford: Oxford University Press.

Holmes, Frederic Lawrence. 2004. *Investigative Pathways: Patterns and Stages in the Careers of Experimental Scientists*. New Haven, CT: Yale University Press.

Hoover, David L. 2016. "Argument, Evidence, and the Limits of Digital Literary Studies." In *Debates in the Digital Humanities 2016*, edited by Matthew K. Gold and Lauren F. Klein, 230–250. Minneapolis, MN: University of Minnesota Press.

Hughes, Robert. 1991/1980. *The Shock of the New: Art and the Century of Change*. London: Thames and Hudson.

Huron, David. 2007. *Sweet Anticipation: Music and the Psychology of Expectation*. Cambridge, MA: MIT Press.

Hutchins, Edwin. 1987. *Metaphors for Interface Design*. ICS Report 8703. La Jolla, CA: Institute for Cognitive Science, University of California, San Diego. https://citeseerx.ist.psu.edu/viewdoc/download?doi=10.1.1.861.2756&rep=rep1&type=pdf.

Hutchins, Edwin. 1995. *Cognition in the Wild*. Cambridge, MA: MIT Press.

Hyde, Lewis. 1998. *Trickster Makes This World: Mischief, Myth, and Art*. New York: Farrar, Straus and Giroux.

Ingold, Tim. 2010. "The Man in the Machine and the Self-Builder." In *History and Human Nature: An essay by G E R Lloyd with invited responses*, edited by Brad Inwood and Willard McCarty. Special issue of *Interdisciplinary Science Reviews* 35.3–4: 353–64.

Ingold, Tim. 2011. *Being Alive: Essays on Movement, Knowledge and Description*. London: Routledge.

Israel, Giorgio, and Ana Millán Gasca. 2009. *The World as Mathematical Game: John von Neumann and Twentieth Century Science*. Translated Ian McGilvray. Basel: Birkhäuser.

Jacob, François. 1998/1997. *Of Flies, Mice, and Men*. Translated Giselle Weiss. Cambridge, MA: Harvard University Press.

Jianhui, Li. 2019. "Transcranial Theory of Mind: A New Revolution of Cognitive Science." *International Journal of Philosophy* 7 (2): 66–71.

Johnson, Curtis. 2015. *Darwin's Dice: The Idea of Chance in the Thought of Charles Darwin*. Oxford: Oxford University Press.

Johnson-Laird, P. N. 2002. "How Jazz Musicians Improvise." *Music Perception* 19 (3): 415–442.

John-Steiner, Vera. 1985. *Notebooks of the Mind: Explorations of Thinking*. Rev. edn. Oxford: Oxford University Press.

Johnston, Sarah Iles. 2001. "Charming Children: The Use of the Child in Ancient Divination." *Arethusa* 34 (1): 97–117.

Johnston, Sarah Iles. 2003. "Lost in the Shuffle: Roman Sortition and Its Discontents." *Archiv für Religionsgeschichte* 5 (1): 146–156.

Johnston, Sarah Iles. 2005. "Delphi and the Dead." In Johnston and Struck 2005: 283–306.

Johnston, Sarah Iles. 2008. *Ancient Greek Divination*. Chichester: Wiley-Blackwell.

Johnston, Sarah Iles, and Peter T. Struck, eds. 2005. *Mantikê: Studies in Ancient Divination*. Leiden: Brill.

Karppi, Tero, and Yvette Granata. 2019. "Non-artificial Non-intelligence: Amazon's Alexa and the Frictions of AI." *AI & Society* 34: 867–876.

Kaufman, James C., and Robert J. Sternberg, eds. 2010. *The Cambridge Handbook of Creativity*. Cambridge: Cambridge University Press.

Kay, Alan Curtis. 1969. *The Reactive Engine*. Unpublished doctoral dissertation, University of Utah.

Kay, Alan C. 1970. *The Reactive Engine*. Unpublished doctoral dissertation, University of Utah. https://www.proquest.com/docview/302356976/2CA0140B4B79477DPQ/4?accountid=11862 (28 November 2019).

Keller, Evelyn Fox. 1996. "The Dilemma of Scientific Subjectivity in a Postvital Culture". In Galison and Stump 1996: 417–27.

Keller, Evelyn Fox. 2002. *Making Sense of Life: Explaining Biological Development with Models, Metaphors, and Machines*. Cambridge, MA: Harvard University Press.

Keller, Evelyn Fox. 2007. "Whole Bodies, Whole Persons? Cultural Studies, Psychoanalysis, and Biology." In *Subjectivity: Ethnographic Investigations*, edited by João Biehl, Byron Good, and Arthur Kleinman, 352–361. Berkeley, CA: University of California Press.

Keller, Evelyn Fox. 2008. "Organisms, Machines, and Thunderstorms: A History of Self-organization, Part One." *Historical Studies in the Natural Sciences* 38 (1): 45–75.

Keller, Evelyn Fox. 2009. "Organisms, Machines, and Thunderstorms: A History of Self-organization, Part two. Complexity, Emergence, and Stable Attractors." *Historical Studies in the Natural Sciences* 39 (1): 1–31.

Keller, Evelyn Fox. 2020. "Cognitive Functions of Metaphor in the Natural Sciences". In *Making Sense of Metaphor: Evelyn Fox Keller and Commentators on Language and Science*, edited by Marga Vicedo and Denis Walsh. *Interdisciplinary Science Reviews* 45: 249–63.

Keynes, John Maynard. 1921. *A Treatise on Probability*. London: Macmillan and Co.

Kim, Yung Sik. 2018. "Chŏng Yak-yong and *Yijing* Divination". In Lackner 2018: 345–65.

Kirsh, David. 2014. "The Importance of Chance and Interactivity in Creativity." *Pragmatics and Cognition* 22 (1): 5–26.

Kirschenbaum, Matthew G. 2008. *Mechanisms: New Media and the Forensic Imagination*. Cambridge, MA: MIT Press.

Kittler, Friedrich A. 2013. *The Truth of the Technological World: Essays on the Genealogy of Presence*. Stanford: Stanford University Press.

Klahr, David. 2000. *Exploring Science: The Cognition and Development of Discovery Processes*. Cambridge, MA: MIT Press.

Klerer, Melvin. 1968. "Interactive Programming and Automated Mathematics". In *Interactive Systems for Experimental Applied Mathematics*, edited by Melvin Klerer and Juris Reinfelds. Proceedings of the Association for Computing Machinery, Inc. Symposium held in Washington, DC, August 1967. New York: Academic Press.

Kleitman, Daniel J. 2000. "On the Future of Combinatorics." In *Essays on the Future in Honor of Nick Metropolis*, edited by Siegfried S. Hecker and Gian-Carlo Rota, 123–134. New York: Springer Science+Business Media.

Knuth, Donald. 1998. *Seminumerical Algorithms. Vol. 2 of The Art of Computer Programming*. 3rd ed. New York: Addison Wesley Longman.

Knuth, Donald, and Luis Trabb Pardo. 1976. "The Early Development of Programming Languages". STAN-CS-76-562. Stanford CA: Computer Science Department, Stanford University.

Kung, Joseph P. S. 1995. *Gian-Carlo Rota on Combinatorics: Introductory Papers and Commentaries*. Boston: Birkhäuser.

Langton, Christopher G. 1986. "Studying Artificial Life with Cellular Automata." *Physica* 22D: 120–149.

Langton, Christopher G. 1992. "Life at the Edge of Chaos." In *Artificial Life II. Proceedings of the Workshop on Artificial Life Held February, 1990, in Santa Fe, New Mexico*, edited by Christopher G. Langton, Charles Taylor, J. Doyne Farmer, and Steen Rasmussen, 41–91. Redwood City, CA: Addison-Wesley.

Lasdun, James. 2020. "Kinks and Convolutions". *London Review of Books* 42.4. https://www.lrb.co.uk/the-paper/v42/n04/james-lasdun/kinks-and-convolutions (18 June 2020).

Latour, Bruno, and Steve Woolgar. 1986/1979. *Laboratory Life: The Construction of Scientific Facts*. Princeton, NJ: Princeton University Press.

Laurel, Brenda. 2014. *Computers as Theatre*. 2nd ed. Upper Saddle River, NJ: Addison-Wesley.

Leary, David E., ed. 1990. *Metaphors in the History of Psychology*. Cambridge: Cambridge University Press.

L'Ecuyer, Pierre. 2004. "Random Number Generation." In *Handbook of Computational Statistics: Concepts and Methods*, edited by James E. Gentle, Wolfgang Härdle, and Yuichi Mori, 35–70. Berlin: Springer.

Lenoir, Timothy. 2007. "Techno-humanism: Requiem for the Cyborg." In *Genesis Redux: Essays in the History and Philosophy of Artificial Life*, edited by Jessica Riskin, 196–220. Chicago: University of Chicago Press.

Levesque, Hector J. 2014. "Our Best Behaviour." *Artificial Intelligence* 212: 27–35.

Levinas, Emmanuel. 1999/1995. *Alterity and Transcendence*. Translated Michael B. Smith. London: Athlone Press.

Lezra, Jacques, and Liza Blake, eds. 2016. *Lucretius and Modernity: Epicurean Encounters Across Time and Disciplines*. Houndmills, Basingstoke: Palgrave.

Lloyd, G. E. R. 1973. *Greek Science after Aristotle*. New York: W. W. Norton & Company.

Lloyd, G. E. R. 1989. *The Revolutions of Wisdom: Studies in the Claims and Practice of Ancient Greek Science*. Berkeley, CA: University of California Press.

Lloyd, G. E. R. 1990. *Demystifying Mentalities*. Cambridge: Cambridge University Press.

Lloyd, G. E. R. 1991. "The Invention of Nature." Chap. 18 in *Methods and Problems in Greek Science: Selected Papers*, 417–434. Cambridge: Cambridge University Press.

Lloyd, G. E. R. 1996. "Comparative Studies and Their Problems: Methodological Preliminaries." Chap. 1 in *Adversaries and Authorities: Investigations into Ancient Greek and Chinese Science*. Cambridge: Cambridge University Press.

Lloyd, G. E. R. 2002. *The Ambitions of Curiosity: Understanding the World in Ancient Greece and China*. Cambridge: Cambridge University Press.

Lloyd, G. E. R. 2019. "The Clash of Ontologies and the Problems of Translation and Mutual Intelligibility." In Lloyd and Vilaça 2019: 36–43.

Lloyd, G. E. R., and Aparecida Vilaça, eds. 2019. "Science in the Forest, Science in the Past." *Special Issue of HAU: Journal of Ethnographic Theory* 9 (1): 36–182.

Long, A. A. 2006. *From Epicurus to Epictetus: Studies in Hellenistic and Roman Philosophy.* Oxford: Clarendon Press.

Lorenz, Edward. 1993. *The Essence of Chaos.* Seattle: University of Washington Press.

Lovász, László. 2005. *Combinatorial Problems and Exercises.* 2nd ed. Providence, RI: American Mathematical Society.

Lovász, László, D. B. Shmoys, and É Tardos. 1995. "Combinatorics in Computer Science". In Graham, Grötschel and Lovász 1995.

Loewe, Michael, and Carmen Blacker, eds. 1981. *Divination and Oracles.* London: George Allen & Unwin.

Luff, Paul, Nigel Gilbert, and David Frolich, eds. 1990. *Computers and Conversation.* London: Academic Press.

Mahoney, Michael Sean. 2011. *Histories of Computing,* edited by Thomas Haigh. Cambridge, MA: Harvard University Press.

March, James G. 2010. "Generating Novelty." Chap. 4 in *The Ambiguities of Experience,* 74–98. Ithaca, NY: Cornell University Press.

Markoš, Anton, Jana Švorcová, and Josef Lhotský. 2017. "Living as Languaging: Distributed Knowledge in Living Beings." In *Cognition Beyond the Brain: Computation, Interactivity and Human Artifice,* edited by Stephen J. Cowley and Frédéric Vallée-Tourangeau, 193–214. 2nd ed. London: Springer-Verlag.

Maurizio, L. 1995. "Anthropology and Spirit Possession: A Reconsideration of the Pythia's Role at Delphi." *Journal of Hellenic Studies* 115: 69–86.

Maxwell, James Clerk. 1864/1855-6. "On Faraday's Lines of Force." *Transactions of the Cambridge Philosophical Society* 10 (1): 27–83.

Maxwell, James Clerk. 1965/1870. "Address to the Mathematical and Physical Sections of the British Association." In *The Scientific Papers of James Clerk Maxwell, Volume Two,* edited by W. D. Niven, 215–229. New York: Dover Publications.

McCarty, Willard. 2014/2005. "Modelling." Chap. 1 in *Humanities Computing,* 20–72. Rev. edn. Houndmills: Palgrave Macmillan.

McCarty, Willard. 2018. "Modelling What There Is: Ontologising in a Multidimensional World". In Ciula, Eide, Marras and Sahle 2018: 33–45.

McCarty, Willard. 2019a. "Modeling the Actual, Simulating the Possible." In *The Shape of Data in the Digital Humanities: Modeling Texts and Text-Based Resources,* edited by Julia Flanders and Fotis Jannidis, 264–284. London: Routledge.

McCarty, Willard. 2019b. "Modelling, Ontology and Wild Thought: Toward an Anthropology of the Artificially Intelligent." In Lloyd and Vilaça 2019: 147–161.

McCarty, Willard. 2019c. "Modeling, Ontology and Wild Thought: Toward an Anthropology of the Artificially Intelligent." In *Science in the Forest. Science in the Past,* edited by Geoffrey E. R. Lloyd and Aparecida Vilaça, 147–161. Special issue of HAU.

Menary, Richard, ed. 2010. *The Extended Mind.* Cambridge, MA: MIT Press.

Merton, Robert K. 1968. *Social Theory and Social Structure.* Enl. ed. New York: The Free Press.

Miller, John H., and Scott E. Page. 2007. *Complex Adaptive Systems: An Introduction to Computational Models of Social Life.* Princeton, NJ: Princeton University Press.

Mirowski, Philip. 2002. *Machine Dreams: Economics Becomes a Cyborg Science.* Cambridge: Cambridge University Press.

Moerman, Michael. 1988. *Talking Culture: Ethnography and Conversation Analysis.* Philadelphia, PA: University of Pennsylvania Press.

Monod, Jacques. 1972/1970. *Chance and Necessity: An Essay on the Natural Philosophy of Modern Biology*. Translated Austryn Wainhouse. London: Collins.

Moore, Omar Khayyam. 1957. "Divination – A New Perspective." *American Anthropologist* 59 (1): 69–74.

Moretti, Franco. 2013. *Distant Reading*. London: Verso.

Morgan, Mary S. 2012. *The World in the Model: How Economists Work and Think*. Cambridge: Cambridge University Press.

Motte, Warren F., Jr. 1986. "Clinamen Redux". *Comparative Literature Studies* 23 (4): 263-281.

Motte, Warren F., Jr., ed. and trans. 1998/1986. *Oulipo: A Primer of Potential Literature*. Normal IL: Dalkey Archive Press.

Mukerji, Chandra. 1998. "The Collective Construction of Scientific Genius." In *Cognition and Communication at Work*, edited by Yrjö Engeström, and David Middleton, 257–278. Cambridge: Cambridge University Press.

Neale, Dennis C., and John M. Carroll. 1997. "The Role of Metaphors in User Interface Design." In *Handbook of Human-Computer Interaction*. edited by M. Helander, T. K. Landauer, and P. Prabhu, 441–462. 2nd ed. Amsterdam: Elsevier Science.

Nersessian, Nancy J. 1987. "A Cognitive-Historical Approach to Meaning in Scientific Theories." In *The Process of Science: Contemporary Philosophical Approaches to Understanding Scientific Practice*, edited by Nancy J. Nersessian, 161–177. Dordrecht: Martinus Nijhoff.

Nersessian, Nancy J. 1992. "How Do Scientists Think? Capturing the Dynamics of Conceptual Change in Science." In *Cogntive Models of Science*, edited by Ronald N. Giere, 3–44. Minneapolis, MN: University of Minnesota Press.

Nersessian, Nancy J. 2002. "Maxwell and 'the Method of Physical Analogy': Model-Based Reasoning, Generic Abstraction, and Conceptual Change." In *Reading Natural Philosophy: Essays in the History and Philosophy of Science and Mathematics*, edited by David B. Malament, 129–166. Chicago, IL: Open Court.

Nersessian, Nancy J. 2008. *Creating Scientific Concepts*. Cambridge, MA: MIT Press.

Netz, Reviel. 2009. *Ludic Proof: Greek Mathematics and the Alexandrian Aesthetic*. Cambridge: Cambridge University Press.

Newman-Norlund, Sarah E., Matthijs L. Noordzij, Roger D. Newman-Norlund, Inge AC Volman, Jan Peter De Ruiter, Peter Hagoort, and Ivan Toni. 2009. "Recipient Design in Tacit Communication." *Cognition* 111: 46–54.

Nicolau, Gaspar Pujol. 2009. "Traditional Cosmological Symbolism in Ancient Board Games." Unpublished doctoral dissertation, Universitat Internacional de Catalunya, Barcelona. https://www.tesisenred.net/handle/10803/387431 (18 June 2020).

Norman, Donald A., and Stephen W. Draper, eds. 1986. *User Centered System Design: New Perspectives on Human-Computer Interaction*. Hillsdale, NJ: Lawrence Erlbaum Associates.

North, Michael. 2013. *Novelty: A History of the New*. Chicago: University of Chicago Press.

O'Keefe, Tim. 2009. "Action and Responsibility." In *The Cambridge Companion to Epicureanism*, edited by James Warren, 142–309. Cambridge: Cambridge University Press.

Osbeck, Lisa M., Nancy J. Nersessian, Kareen R. Malone, and Wendy C. Newstetter. 2011. *Science as Psychology: Sense-Making and Identity in Science Practice*. Cambridge: Cambridge University Press.

Pais, Alexandre. 2013. "Ethnomathematics and the Limits of Culture." *For the Learning of Mathematics* 33 (3): 2–6.

Parhami, Behrooz. 2003. "Number Representation and Computer Arithmetic." In *Encyclopedia of Information Systems*, Volume 3, edited by Hossein Bidgoli, 317–333. Amsterdam: Academic Press.

Park, George K. 1963. "Divination and Its Social Contexts." *Journal of the Royal Anthropological Institute of Great Britain and Ireland* 93 (2): 195–209.

Pask, Gordon. 1961. *An Approach to Cybernetics.* London: Hutchinson.

Pask, Gordon. 1971. "A Comment, a Case History and a Plan". In Reichardt 1971: 76–110.

Peek, Phillip M. 1991. "The Study of Divination, Present and Past." In *African Divination Systems: Ways of Knowing,* edited by Phillip M. Peek, 1–22. Bloomington, IN: Indiana University Press.

Perez-Marin, Diana, and Ismael Pascual-Nieto. 2011. *Conversational Agents and Natural Language Interaction: Techniques and Effective Practices.* Hershey, PA: Information Science Reference.

Perry, Mark. 2017. "Socially Distributed Cognition in Loosely Coupled Systems." In *Cognition Beyond the Brain: Computation. Interactivity and Human Artifice,* edited by Stephen J. Cowley and Frédéric Vallée-Tourangeux, 19–41. Cham, Switzerland: Springer International.

Pickering, Andrew. 2010. *The Cybernetic Brain: Sketches of Another Future.* Chicago: University of Chicago Press.

Poincaré, Henri. 1914/1908. *Science and Method.* Translated and edited by Francis Maitland. London: Thomas Nelson and Sons.

Quine, Willard Vn Orman. 2013. *Word and Object.* New ed. Cambridge, MA: MIT Pres.

Raphals, Lisa. 2013. *Divination and Prediction in Early China and Ancient Greece.* Cambridge: Cambridge University Press.

Reichardt, Jasia. 1968. *Cybernetic Serendipity.* London: Studio International.

Reichardt, Jasia. 1971. *Cybernetics, Art and Ideas.* London: Studio Vista.

Reichman (Adar), Rachel. 1986. "Communication Paradigms for a Windows System." In *User Centered System Design: New Perspectives on Human-Computer Interaction,* edited by Donald A. Norman and Stephen W. Draper, 285–313. Hillsdale, NJ: Lawrence Erlbaum Associates.

Rheinberger, Hans-Jörg. 2010. *An Epistemology of the Concrete: Twentieth-Century Histories of Life.* Durham, NC: Duke University Press.

Rochberg, Francesca. 2016. *Before Nature: Cuneiform Knowledge and the History of Science.* Chicago, IL: University of Chicago Press.

Rorty, Richard. 2004. "Being That Can Be Understood Is Language." In *Gadamer's Representations: Reconsidering Philosophical Hermeneutics,* edited by Markus Krajewski, 21–29. Berkeley, CA: University of California Press.

Ross, Lee, Mark Lepper, and Andrew Ward. 2010. "History of Social Psychology: Insights, Challenges, and Contributions to Theory and Application." In *Handbook of Social Psychology,* Vol 1, edited by Susan T. Fiske, Daniel T. Gilbert, and Gardiner Lindzey, 3–50. Hoboken, NJ: John Wiley.

Sawyer, R. Keith. 2003. *Improvised Dialogues: Emergence and Creativity in Conversation.* Westport, CN: Ablex.

Schegloff, Emmanuel A. 1991. "Conversation Analysis and Socially Shared Cognition." In *Perspectives on Socially Shared Cognition,* edited by L. Resnick, J. Levine, and S. Teasley, 150–171. Washington, DC: American Psychological Association.

Schiffrin, Deborah, Deborah Tannen, and Heidi E. Hamilton, eds. 2001. *The Handbook of Discourse Analysis.* Oxford: Blackwell.

Schütz, Alfred. 1951. "Making Music Together: A Study in Social Relationship." *Social Research* 18 (1): 76–97.

Schulz, Bruno. 1998. *The Collected Works of Bruno Schulz,* edited by Jerzy Ficowski. London: Picador.

Seife, Charles. 2014. "Randomness." In *The Best Writing in Mathematics 2013*, edited by Mircea Pitici, 52–55. Princeton, NJ: Princeton University Press.

Selin, Helaine, ed. 2016. *Encyclopaedia of the History of Science, Technology, and Medicine in Non-Western Cultures*. 3rd ed. Dordrecht: Springer Science+Business Media.

Shermer, Michael. 1995. "Exorcising Laplace's Demon: Chaos and Antichaos, History and Metahistory." *History and Theory* 34 (1): 59–83.

Shiryaev, A. N. 2000. "Andreĭ Nikolaevich Kolmogorov (April 25m 1903 to October 20, 1987). A Biographical Sketch of his Life and Creative Paths." In *Kolmogorov in Perespective*, Translated by Harold H. McFaden, 1–88. Washington, DC: American Mathematical Society and London Mathematical Society.

Shklovsky, Viktor. 1965/1917. "Art as Technique." In *Russian Formalist Criticism: Four Essays*, edited by Lee T. Lemon, and Marion J. Reis, 3–24. Lincoln, NB: University of Nebraska Press.

Shneiderman, Ben. 1980. *Software Psychology: Human Factors in Computer and Information Systems*. Cambridge, MA: Winthrop Publishers.

Sidnell, Jack, and Tanya Stivers. 2013. *The Handbook of Conversation Analysis*. Chichester, West Sussex: Wiley-Blackwell.

Simon, Herbert A. 1986/1983. "Understanding the Processes of Science: The Psychology of Scientific Discovery". In *Progress in Science and Its Social Conditions*. Nobel Symposium 58 held at Lidingö, Sweden, 15-19 August 1983. Oxford: Pergamon Press.

Simmel, Georg. 2009/1908. *Sociology: Inquiries into the Construction of Social Forms*. Vol. 1. Translated and edited by Anthony J. Blasi, Anton K. Jacobs and Matthew Kanjirathinkal. Leiden: Brill.

Simonton, Dean Keith. 2011. "Creativity and Discovery as Blind Variation: Campbell's (1960) BVSR Model After the Half-Century Mark." *Review of General Psychology* 15 (2): 158–174.

Smith, Richard J. 1991. *Fortune-tellers and Philosophers: Divination in Traditional Chinese Society*. Boulder, CO: Westview Press.

Smith, Leonard A. 2007. *Chaos: A Very Short Introduction*. Oxford: Oxford University Press.

Soare, Robert I. 2009. "Turing's Oracle Machines, Online Computing, and Three Displacements in Computability Theory." *Annals of Pure and Applied Logic* 160: 368–399.

Star, Susan Leigh. 1998. "Working Together: Symbolic Interactionism, Activity Theory, and Information Systems." In *Cognition and Communication at Work*, edited by Yrjö Engeström, and David Middleton, 296–318. Cambridge: Cambridge University Press.

Starr, Ivan. 1990. *Queries to the Sun God: Divination and Politics in Sargonid Assyria*. State Archives of Assyria, Volume IV. Helsinki: Helsinki University Press.

Steinle, Friedrich. 2016/2005. *Exploratory Experiments: Ampère, Faraday, and the Origins of Electrodynamics*. Translated by Alex Levine. Pittsburgh PA: University of Pittsburgh Press.

Stewart, Ian. 1989. *Does God Play Dice? The Mathematics of Chaos*. Cambridge, MA: Blackwell.

Strathern, Marilyn. 1988. *The Gender of the Gift: Problems with Women and Problems with Society in Melanesia*. Berkeley, CA: University of California Press.

Strathern, Marilyn. 1999. *Property, Substance and Effect: Anthropological Essays on Persons and Things*. London: Athalone Press.

Strathern, Marilyn. 2005. *Partial Connections*. Rev. ed. Walnut Creek, CA: AltaMira Press.

Strathern, Marilyn. 2020. *Relations: An Anthropological Account*. Durham, NC: Duke University Press.

Streeck, Jürgen. 1995. "On Projection". In Goody 1995: 87–110.

Struck, Peter T. 2016. *Divination and Human Nature: A Cognitive History of Intuition in Classical Antiquity*. Princeton, NJ: Princeton University Press.

Suchman, Lucy. 1998. "Human/Machine Reconsidered." *Cognitive Studies* 5 (1): 5–13.

Suchman, Lucy. 2007. *Human-Machine Reconfigurations. Plans and Situated Actions*. 2nd ed. Cambridge: Cambridge University Press.

Sudnow, David. 2001. *Ways of the Hand: A Rewritten Account*. Cambridge, MA: MIT Press.

Swain, Merrill. 2006. "Languaging, Agency and Collaboration in Advanced Second Language Proficiency." In *Advanced Language Learning: The Contribution of Halliday and Vygotsky*, edited by Heidi Byrnes, 95–108. London: Continuum.

Taruskin, Richard. 2003. "Stravinsky and Us." In *The Cambridge Companion to Stravinsky*, edited by Jonathan Cross, 260–284. Cambridge: Cambridge University Press.

Taussig, Michael. 1993. *Mimesis and Alterity: A Particular History of the Senses*. New York: Routledge.

te Molder, Hedwig, and Jonathan Potter, eds. 2005. *Conversation and Cognition*. Cambridge: Cambridge University Press.

Thom, René. 1980. "*Halte au hazard, silence au* bruit". *Le Débat* 3: 1-10. Trans. "Stop chance! Silence noise!" by Robert E. Chumbley, *SubStance* 12.3 (1983), issue 40: 11–21.

Thomas, Peter J., ed. 1995. *The Social and Interactional Dimensions of Human-Computer Interfaces*. Cambridge: Cambridge University Press.

Thompson, Michael. 1974. "Intelligent Computers and Visual Artists." *Leonardo* 7 (3): 227–234.

Tudini, Vincenza, and Anthony J. Liddicoat. 2017. "Computer-mediated Communication and Conversation Analysis." In *Language, Education and Technology*, edited by Steven L. Thorne and Stephen May, 415–426. Cham, Switzerland: Springer International.

Turing, A. M. 2012/1938. *Alan Turing's Systems of Logic: The Princeton Thesis*, edited by Andrew W. Appel with essays by Appel and Solomon Feferman. Princeton, NJ: Princeton University Press.

Turing, A. M. 1950. "Computing Machinery and Intelligence." *Mind; A Quarterly Review of Psychology and Philosophy* 59 (236): 433–460.

Turnbull, William. 2003. *Language in Action: Psychological Models of Conversation*. Hove: Psychology Press.

Turner, Raymond, and Nicola Angius. 2020. "The Philosophy of Computer Science". *Stanford Encyclopedia of Computer Science* (Spring 2020 ed.). Ed. Edward N. Zalta. https://plato.stanford.edu/entries/computer-science/ (17 June 2020).

Tweney, Ryan D. 2002. "Epistemic Artifacts: Michael Faraday's Search for the Optical Effects of Gold." In *Model-Based Reasoning: Science, Technology, Values*, edited by Lorenzo Magnani and Nancy J. Nersessian, 287–303. New York: Springer Science +Business Media.

Tweney, Ryan D. 2013. "Cognitive-Historical Approaches to the Understanding of Science". In Feist and Gorman 2013: 71–93.

Vernant, Jean-Pierre. 1991/1974. "Speech and Mute Signs." In *Mortals and Immortals: Collected Essays*, edited by Froma I. Zeitlin, 303–317. Princeton, NJ: Princeton University Press.

Weaver, Warren. 1949. "The Mathematics of Communication." *Scientific American* 181 (1): 11–15.

von Neumann, John. 1945. *First Draft of a Report on the EDVAC*. Contract No. W-67C-ORD-4926, Between the United States Army Ordnance Department and the University of Pennsylvania. Philadelphia PA: Moore School of Electrical Engineering, University of Pennsylvania.

von Neumann, John. 1951. "The General and Logical Theory of Automata." In *Cerebral Mechanisms of Behavior: The Hixon Symposium*, edited by Lloyd A. Jeffress, 1–41. New York: John Wiley & Sons.

von Neumann, John. 1958. *The Computer and the Brain*. Mrs. Hepsa Ely Silliman Memorial Lectures. New Haven CN: Yale University Press.

Wagner, Roy. 1996. *The Invention of Culture*. 2nd ed. Chicago: University of Chicago Press.

Wagner, Andreas. 2012. "The Role of Randomness in Darwinian Evolution." *Philosophy of Science* 79: 95–119.

Waldrop, M. Mitchell. 1992. *Complexity: The Emerging Science at the Edge of Order and Chaos*. New York: Simon and Schuster.

Watson, David. 2019. "The Rhetoric and Reality of Anthropomorphism in Artificial Intelligence." *Minds and Machines* 29: 417–440.

Weaver, Warren. 1949. "The Mathematics of Communication." *Scientific American* 181 (1): 11–15.

Wegner, Peter. 1998. "Interactive Foundations of Computing." *Theoretical Computer Science* 192: 315–351.

Weizenbaum, Joseph. 1976. *Computer Power and Human Reason: From Judgment to Calculation*. San Francisco: W. H. Freeman.

Wells, Richard B. 2007. "Meanings-Based Networks: A New Learning Paradigm for ART Network Systems Models". LCNTR Tech Brief. http://www.mrc.uidaho.edu/~rwells/techdocs/ (17 June 2020).

Whiten, Andrew. 1999. "Machiavellian Intelligence Hypothesis." In *The MIT Encyclopedia of the Cognitive Sciences*, edited by Robert A. Wilson and Frank C. Keil, 495–496. Cambridge, MA: MIT Press.

Wilson, Robert A., and Lucia Foglia. 2017. "Embodied Cognition". *Stanford Encyclopedia of Philosophy*. Spring 2017 ed. https://plato.stanford.edu/archives/spr2017/entries/embodied-cognition/ (10/4/19).

Wilson, Robin, and John J. Watkins, eds. 2013. *Combinatorics: Ancient and Modern*. Oxford: Oxford University Press.

Winch, Peter. 1959. *The Idea of a Social Science and Its Relation to Philosophy*. 2nd ed. London: Routledge.

Wisdom, J., J. L. Austin, and A. J. Ayer. 1946. "Symposium: Other Minds." *Proceedings of the Aristotelian Society*, Suppl. 20 (Logic and Reality), 122–197.

Winograd, Terry. 2006. "Shifting Viewpoints: Artificial Intelligence and Human-Computer Interaction." *Artificial Intelligence* 170: 1256–1258.

Woodruff, Allison, and Paul M. Aoki. 2004. Conversation Analysis and User Experience." *Digital Creativity* 15 (4): 232–238.

Yon, Daniel, Floris P. de Lange and Clare Press. 2018. "The Predictive Brain as Stubborn Scientist". *Trends in Cognitive Sciences* 23 (1): 6-8.

Zeitlyn, David. 1995. "Divination as Dialogue: Negotiation of Meaning with Random Responses." In Goody 1995: 189–205.

Zeitlyn, David. 2012. "Divinatory Logics: Diagnoses and Predictions Mediating Outcomes." *Current Anthropology* 53 (5): 525–546.

Zuesse, Evan M. 1987. "Divination." In *The Encyclopedia of Religion*, edited by Mircea Eliade, 375–382. London: Macmillan.

Inventing Artificial Intelligence in Ethiopia

Alan F. Blackwell, Addisu Damena and Tesfa Tegegne

ABSTRACT

Artificial Intelligence (AI) research has always been embedded in complex networks of cultural imagination, corporate business, and sociopolitical power relations. The great majority of AI research around the world, and almost all commentary on that research, assumes that the imagination, business, and political systems of Western culture and the Global North are sufficient to understand how this technology should develop in future. This article investigates the context within which AI research is imagined and conducted in the Amhara region of Ethiopia, with implications for public policy, technology strategy, future research in development contexts, and the principles that might be applied as practical engineering priorities.

This is a sequel to the paper on 'Ethnographic Artificial Intelligence' that was presented at the second Cambridge workshop on *Science in the Forest, Science in the Past* (Blackwell 2019). The goal of this programme of research, as presented in that paper, has been to offer a new cultural critique of the field of Artificial Intelligence (AI) that steps away from the investment priorities and social anxieties of the wealthy nations where AI research is typically concentrated.

The original paper proposed an ethnographic programme of research, noting that AI research does not usually employ ethnographic methods. This new contribution has been written from 'the field' in Ethiopia, with two Ethiopian AI researchers joining the original author to report on the findings emerging from the project. It is hardly necessary to observe that the findings of an ethnographic research project are expected to be very different from those anticipated when the project starts – indeed, if they were not different, there would be no point in conducting ethnographic research at all!

One consequence is that this sequel must start by admitting the misunderstandings and errors of the original proposal. We will try to keep this apology short, in order to move on to the more interesting question of what is being discovered. The remainder of the paper then turns to the distinctive cultural, political and economic context of the Amharic-speaking people within the ethnically federal state of Ethiopia, and the question of what might be distinctive about AI research when conducted in this place. Some of the central concerns emerging from this analysis relate firstly to the fact that AI is a technology of the imagination, and secondly that much AI research presumes 'intelligence' will be manifested in ways that are universal among all people. We therefore ask what kinds of technological imagination are appropriate and attractive to Amharic students and researchers, and how these compare to the assumptions of AI as presented in the canonical research literature of AI within computer science.

The agenda of ethnographic AI

Blackwell's original paper offered a contextual critique of AI by drawing on the familiar example of von Kempelen's 'Mechanical Turk' – an automaton that fascinated audiences in the eighteenth Century by playing an expert game of chess (with the aid of an expert player hidden inside) (Schaffer 1999). Blackwell argued that human actors also appear 'inside' artificial constructs in other cultural settings, for example in the *egwugwu* puppets of the Nigerian Igbo (Achebe 1958/2010), and that the form of the resulting AI performances is shaped by the specific sociotechnical imaginaries (cf. Jasanoff and Kim 2015) of the cultures in which they are created. He observed that there was no reason to give priority to the early modern imagination of enlightenment Europe over that of early colonial West Africa, and also that some common dynamics might be identified in the comparison. In particular, all of these 'artificial intelligences' offer a kind of theatre that presents some essence of human agency imaginatively re-embodied in a constructed costume.

Blackwell argued that contemporary AI technology (including the new 'Mechanical Turk' created by the Amazon company – (Irani and Six Silberman 2013)) continues to be engaged in the exercise of abstracting and re-embodying human intelligence in more or less elaborate costumes, through the economic enterprise of big data and machine learning. The ethnographic proposal was to observe alternative essences of mechanical reasoning as they might be conceived in 'Africa', noting that even supposedly universal computations of basic arithmetic may be different or even more effective when expressed in a language such as Yoruba (Verran 2001).

The most embarrassing error in that original argument, when writing this sequel in Bahir Dar – the capital of Amhara – was the assumption that 'Africa' would offer a single analytic frame within which observations of Igbo and Yoruba culture could have any relevance at all to work being done in Bahir Dar. It was easy enough, as an AI engineer working in the University of Cambridge, to observe that people in 'Africa' might see AI differently. And indeed, they do, as discussed in the remainder of this analysis. However, most of the argument in the original paper, while being (possibly) relevant to

AI research in Nigeria, must be discarded in the present work. Having used the African continent as a rhetorical device to gain some critical distance from the supposedly 'universal' perspective of Western computer science (Blackwell 2010), it is now necessary to be far more specific about the history and culture of the Amhara region.

Scholarship and technology in Amhara

Ethiopia draws with pride on one of the oldest continuous literary and cultural traditions in the world. The Ethiopian Orthodox Church has a rich and distinctive body of architecture, liturgical music, visual arts and especially the body of scholarship recorded in the Ge'ez language. Modern Amharic is directly descended from classical Ge'ez, and is still written in the Ge'ez script. Ge'ez has been preserved through a respected monastic tradition, and continues to be taught in church schools today. For centuries, the emperors of Ethiopia and its people have traced descent from the tribe of Judah, and much of the national symbolism draws on corresponding biblical imagery and Semitic languages, rather than identifying with the continent of sub-Saharan Africa.

This heritage includes philosophical, mathematical and astronomical texts that might well be imagined as antecedents of AI. As with any codified and preserved system of knowledge, these texts are linguistic technologies that become embodied within sociotechnical practices. The Ge'ez calendar, for example, encapsulates in the names of each month the understanding of the agricultural work that should be done in that month. To the AI researcher, this calendar can be interpreted as a formal model, inferred from observation, and used to guide action, just as would be expected for computational machine learning systems. Ethiopian scholars note various ways in which properties of computer systems such as binary arithmetic have been anticipated in Ge'ez texts.

National pride in the scholarly traditions of the Orthodox Church is further enhanced by archaeological evidence that locates the origins of humanity within the borders of modern Ethiopia. The famous *Australopithecus* skeleton 'Lucy' is displayed in the National Museum of Ethiopia, alongside many other skeletal remains and stone artefacts demonstrating that the earliest evidence of human culture can be found here. Devout Ethiopian Orthodox Christians will readily observe that Ethiopia must have been the location of the original Paradise, and even suggest that Ge'ez must have been the language of Adam.

Despite this inherited wealth of ancient knowledge, Ethiopia today is not a wealthy country. Investment in technology arrives sporadically, and in inconsistent ways. For example, many Ethiopians comment on the great quantity of investment from the People's Republic of China, highly visible in the construction industry. Such investments can result in serious distortions to the local economy. The eucalyptus timber that was introduced a century ago to build the walls of traditional dwellings is instead taken to the cities as scaffolding for construction of multi-storey concrete buildings.

Concrete reinforcing rod is readily available, but in the Bahir Dar market, a black-smith breaks this raw material into pieces to be hammered into axes, chisels and tra-ditional iron ploughs for use by local farmers with their oxen.

Despite relying on pre-industrial farming methods, and tools that are hand-made from iron and wood, young people in the rural communities of Amhara are very much alert to the value of scientific education in addressing the chal-lenges of their families. Rural schoolchildren say that they want to be water engineers or veterinary doctors – addressing the most significant and intract-able problems of rural subsistence farming. In Bahir Dar and other cities, the value of information and communication technologies (ICT) for efficient urban life is similarly salient. Students of computer science see many opportu-nities for ICT to address health challenges, to make agricultural markets more efficient, or direct resources more efficiently toward the housing and transport problems of city residents.

Yet in ICT, just as in the construction industry, opportunities for local inno-vation are dependent on infrastructure investment priorities. Electronic media in Ethiopia have developed within a policy framework that struggles to prior-itize popular engagement and debate, given an entrenched legacy of internal dissent in the wake of the Marxist-Leninist military junta of the 1970s and 80s. A more compatible partner for the state telecom company was Chinese company ZTE, whose 2006 investment was the largest in the history of telecom-munications in Africa (Gagliardone 2016). The use of media in governance is exemplified by Woredanet, an initiative that broadcasts central directives to regional government offices rather than promoting decentralization or local engagement (Gagliardone 2014).

Nevertheless, as in many other developing countries, the cellular telephone network has been deployed far more rapidly than the wired internet, with the added advantage that mobile phones are more robust to intermittent fail-ures of electrical power supply. But Ethiopia has continued to score at the bottom of regional and global rankings in terms of access to ICTs, and there is very little local software industry. Smartphone users expect operating systems and utility apps to be free, and there is little profit opportunity or investment in local language apps, beyond government support for national industries. Users of personal computers have an asymmetric relationship with multinational software companies, dependent on 'cracked' free versions of popular applications, but with talented programmers seeking jobs and investment from such companies in order to pay their salaries while they devote their free time to Christian charity and public service.

Artificial intelligence in Bahir Dar

How might the distinctive context of Ethiopia and Amhara shape research into AI? Bahir Dar University, where the Bahir Dar Institute of Technology was the first national centre of technology research, has a strong AI research group led from the ICT4D (Information and Communication Technologies for Development) Center. The focus of this group is specifically on the application of AI technologies for health, social and economic development.

Much of this group's research is focused on the application of natural language processing (NLP) methods. Typical research problems in NLP include speech understanding (deriving text from spoken audio), speech generation from text, document summarization, retrieval of information on the basis of a text query, answering questions, or translation between languages. At first sight, many of these problems might appear to Western readers to be solved already – 'voice assistant' products such as Amazon's Alexa, Apple Siri, or Google Assistant routinely do all of these things. However, two classical definitions of AI give cause for hesitation. The first is that, if these current techniques were able to maintain continuous natural conversations, we would be surprisingly close to meeting the criteria of the Turing Test – a development that seems very distant, both from critical analysis of the test itself (Collins 2018) and from the everyday experience of people using such products, which are often disappointingly stupid in practice. The second reason for hesitation is a perennial problem in all areas of AI research – that the central and exciting role of imagination in AI dissipates once a concrete invention has actually been achieved. This characteristic of AI as always-imaginary led to the tongue-in-cheek definition from the MIT AI Lab in the 1980s that 'if it works, it isn't AI'.

So, in practice, the NLP technologies that are the focus of the ICT4D group in Bahir Dar concentrate primarily on creating the same kind of basic behaviour that is also being routinely deployed by companies like Apple, Amazon and Google – question-answering, speech synthesis and speech recognition. There are two important challenges in such research, one greater than the other. The first is to replicate the advances that have already been made in processing the English language, but applying the same techniques instead to Amharic. Amharic does have interesting and distinctive linguistic features, not least the phonetics of the Ge'ez script. This does involve a number of practical aspects that must be approached differently from standard methods.

But the greatest technical challenge is one that Amharic shares with many other languages of the world, which is that it is a 'low resource' language. NLP research, especially using machine learning methods, relies on large corpora of natural language text extracted from newspaper archives, dictionaries, literature, etc., and aggregated into research collections such as the British National Corpus. These annotated corpora are used to train statistical language models that recognize common features such as part-of-speech, word morphology, sentence structure, pronunciation, prosody and so on. In a low-resource language, each of these components of an NLP system must be reconstructed, often using much smaller data sets than were previously used to achieve state-of-the-art performance benchmarks in English. The problems of adapting research technologies originally constructed in English seem to recapitulate those of national digital initiatives such as the Schoolnet programme, that delivers educational content via large screens in schools across the whole country, but in a standardized English that can impede comprehension by students in regional language communities (Gagliardone 2014, 291).

Although there is a substantial disparity of data resources between academic research in English and academic research in low-resource languages, there is further and increasing disparity between the resources available to academic researchers in Western universities, and those of commercial laboratories such as Google or Facebook, whose parent companies have access to massive data sets contributed directly by the customers in their global monopoly businesses. Differences in the size of training data sets are particularly salient when 'deep learning' approaches are being used, because these methods require far larger quantities of data to extract the features and patterns for their statistical language models. The current vogue for deep learning research is highlighting the disparities in data resources between academic research in English versus Amharic, and the even greater disparity between commercial research in English, and academic research in Amharic.

The work of the AI group in Bahir Dar includes substantial effort dedicated to the adaptation of NLP techniques from US, UK and corporate research groups, all of which carry out their primary research in English. Researchers in the Bahir Dar group must either create their own data sets in Amharic, or else identify alternative strategies that are suited to low resource languages. Generally, the benchmark of achievement for their results, as with work in other low resource languages, is to see how closely the Amharic results approach those already achieved in English. Unfortunately, where very large English data sets have already been used with deep learning methods, it is unlikely that the results in Amharic will be technically superior to results previously achieved in English, especially if the algorithms used are the same ones that have been developed for, and proved effective with, the English language.

Nevertheless, effective NLP systems could be particularly valuable for the Amharic-speaking population of Ethiopia. Literacy rates in the Amhara region are low, and voice interfaces offer the potential for low-literacy individuals to access digital services for health, finance, agricultural expertise and other knowledge resources by speaking to computers rather than reading and writing. Research students in the ICT4D group at Bahir Dar aspire to deliver dialog-based systems of these kinds, as a public service and to improve the lives of their local population.

However, the disparity between practical value and research advances can be dispiriting for the AI researchers of Ethiopia. They do have an opportunity to apply their skills in ways that may benefit Ethiopian people, but the more reliable methods (that might routinely be applied to commercial products in rich countries) often seem unimpressive or outdated to laboratory researchers in those countries who use huge English data sets to experiment with deep learning methods. Academic assessment in Ethiopia, like other countries, rewards publications in high impact journals and competitive conferences. Results that simply replicate earlier results (even if applied to a new language) will not be considered sufficiently innovative for publication in the most prestigious venues.

Younger students, who have not yet carried out such experiments themselves, imagine that AI might also be the key to address the many technical and economic challenges faced by Ethiopian farmers. But as they learn more about the business-oriented research from the laboratories of Amazon, Google, Uber and others, they realize that Western research fashions have little relevance to the challenges of agriculture. The few Western research venues that do combine AI and agricultural research focus on applications such as greenhouse automation, GPS-guided tractors, or robot fruit-pickers – all projects that seem scarcely relevant in a countryside where there are no tractors, no greenhouses, and far more people than fruit trees.

One talented student interviewed at the ICT4D centre described the choice she faced as follows: 'you can choose either to solve local problems, or to do research'. Like many educated people in Ethiopia, she is committed to using her skills to address local problems – but she does not expect this commitment will be a feature of her professional career. Based on her academic qualifications, she might find a salaried job as a software developer, perhaps for a multinational corporation, which would give her the freedom to contribute to more practical problems on a voluntary basis in her own time. Knowledge-based institutions in Ethiopia, such as local hospitals, also struggle to understand why academics are not interested in addressing local problems in their research. They see this as a social deficiency that must be characteristic of African countries – but don't universities everywhere in the world fail to make connections between research and local problems?

Pasteur's quadrant in Ethiopia

Why might Ethiopian students (or students in any country) believe that solving local problems is not compatible with research? A valuable perspective on this problematic dichotomy comes from the science policy analysis of Donald Stokes, whose book 'Pasteur's Quadrant' proposes an alternative characterization (Stokes 1997). Stokes sets out to challenge the assumption that scientific research can be classified on a simple axis from pure to applied – either seeking fundamental understanding for its own sake, or else motivated by the need to solve immediate problems. The Ethiopian student who tells us that 'you can choose either to solve local problems or to do research' is repeating precisely this choice between fundamental understanding versus immediate problems,

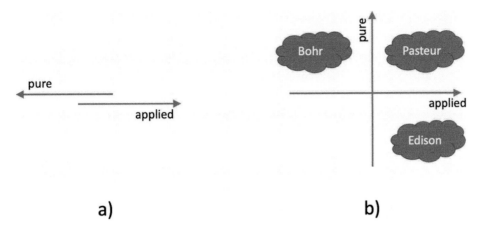

Figure 1. Pure and applied science, as characterized by Stokes (1997). Rather than being located at opposite ends of a continuum (a), pure and applied can be considered as orthogonal axes (b) in which different quadrants represent alternative combinations of research motivation.

which seems just as salient in Bahir Dar as it did to Stokes when carrying out his research into science policy in the USA.

But Stokes proposes a third alternative, in which research and application are not two extremes of a single continuum (Figure 1(a)), but intersecting orthogonal axes incorporating a sector that he describes as 'Pasteur's Quadrant' (Figure 1(b)). He contrasts the research style of Louis Pasteur with two other quadrants. The first is the search for fundamental understanding – exemplified by Niels Bohr, whose curiosity about the structure of the atom was purely intellectual. A second quadrant presents Thomas Edison as an archetypal example of the practical application of science, a practical inventor whose efforts are solely directed toward working engineering solutions. But rather than accepting Bohr and Edison as opposite extremes of a single continuum from pure to applied, Stokes suggests a third alternative, which he illustrates with the case of Louis Pasteur, and calls 'use-inspired basic research'. Pasteur's investigations of germ theory were motivated by practical problems of disease infection and fermentation, but his discoveries also revolutionized the understanding of microorganisms, leading to breakthroughs in vaccination and pasteurization.

Is there any analogy to Pasteur's Quadrant in AI, and might this offer an alternative framing of AI research in Ethiopia? One important question might be to ask what is the pure research or 'fundamental understanding' sought in any kind of AI research? Is AI fundamentally concerned with understanding of humans? If so, is the fundamental understanding of humans necessarily universal? Will such understanding be the same wherever it is investigated, regardless of who the humans are, or of what culture they have inherited, or what their economic and political circumstances might be? Such attempted universalism seems extremely unwise, despite the AI reliance on supposedly universal principles of cognitive science (critiqued rather comprehensively by Geoffrey Lloyd in his book *Cognitive Variations* (Lloyd 2007)).

There seems ample potential here for new insights to be achieved, through use-inspired enquiry that starts from the Ethiopian context. Potentially, such new insights resemble the innovative perspectives that arise in interdisciplinary research (Wilson and Blackwell 2013). New theoretical discoveries come about, not by providing new answers to the same old questions, but rather from asking new questions.

But if an academic attempts to redefine the questions that define her field, how is she to publish, and how could her research proposals be considered as a priority for funding? Publishers and funders are accustomed to addressing the problems of rich countries. The attention economy logic of citation metrics and impact factors is that those countries where academics have ample time to read each other's work will also be the countries where the research questions are understood by consensus to be 'the best' questions, as determined by applying those metrics. Although African scientists do have access to local publication venues, which may be more likely to address local problems and ask local research questions, current systems of evaluating science will never allow such local venues to be recognized as equally valuable when compared to the attention invested by those in more leisured wealthy countries.

Imagining AI technology in Ethiopia

We have already suggested that AI is a technology of the imagination. One flippant (but not wholly inaccurate) way to define AI is that it is the branch of computer science where we try to make computers work the way they do in the movies. Indeed, many computer science students are inspired by particular science fiction movies they have seen, and devote years of research to recreating a fictional computer. A classic example is the transparent interactive projection in the movie *Minority Report*, which (although it seems somewhat impractical even in the movie) was often invoked by students at the time to explain why we should build systems where computer users wave their hands in the air rather than using keyboards or mice.

Noting the importance of imagination to perception of AI technologies among the general public, substantial research projects have been established to gain better understanding of the impact AI is having on the world, through closer reading of the science fiction books and movies that frame our understanding of technology. The Royal Society, with the Leverhulme Centre for the Future of Intelligence, recently published a report explaining why the narratives of AI are significant to understanding AI itself (Cave et al. 2018). Literary scholar Rachel Adams analyses in detail how the fantasies and morality tales of fiction from E.T.H. Hoffman's 1816 story *The Sandman* to Fritz Lang's 1927 film *Metropolis* and contemporary science fiction films such as Alex Garland's *Ex Machina* and Spike Jonze's *Her* prefigure the built-in stereotypes and gendered biases of AI technology products such as voice assistants Siri and Alexa.

It might even be argued that AI is in fact a branch of literature, rather than a branch of science. As with creative writing, the starting point for an AI project is a leap of imagination – imagining a way in which computers could be made to behave differently. The day-to-day mechanics of constructing AI is also remarkably literary – AI programs, like any software source code, are texts, typed and arranged from a computer keyboard (Colburn 1999; Cox and McLean 2013). And the reception of AI systems, as with the reception of literature, involves reading our screens and mobile devices in relation to systems of cultural expectation and understanding. When considering how ICTs come into existence in Ethiopia and other developing countries, Gagliardone notes that technology often starts as a kind of rhetoric and discourse, and varies from context to context (2016, 17). Bowman provides a valuable case study of this sociotechnical imaginary in the utopian ICT policies of Rwandan president Paul Kagame (Bowman 2015).

What could we imagine to be different, if AI was Ethiopian? Technology can certainly be re-imagined, when writing science fiction. For example, science fiction writers in other African countries have used the trope of alien arrival to reflect on an alternative to colonialism, in which advanced technologies become integrated into African reality, as in Tade Thompson's *Rosewater* series. Neill Blomkamp's South African movie *District 9* reflects the logic of apartheid with disempowered aliens interned in concentration camps. The flipside of colonialism and segregation is the diaspora, and African science fiction also imagines scenarios in which expatriate scientists return or unite to apply their talents in their home country, which provides the central plot device of *Nigerians in Space* by Deji Bryce Olukotun.

The Ethiopian thriller *Dertogada* by Worku (2009, tr. 2012) is a fantasy dedicated to the memory of Ethiopia's most famous scientist, NASA employee Kitaw Ejigu. While the real Kitaw died without returning from the USA, the character in *Dertogada* survives to be struck by conscience 'While I was putting all my efforts on the luxury of space research, my people were starving to death. Is there anything more meaningless than this?' In the novel, Kitaw returns to Ethiopia to build secret underground laboratories, equipped with powerful computers and high-tech gadgets, and a staff of other Ethiopian scientists smuggled back from the countries that were exploiting their talents. The laboratory is hidden within one of the island monasteries that also harbour books of traditional medicine, philosophy and secret wisdom written in Ge'ez. The action hero discovering this hidden wealth is a monk-turned-doctor who integrates traditional wisdom with James Bond-style wits and sexual escapades. His adventures are presented as advancing Ethiopia's position in an international race against not only the rich countries, but countries like

India that have achieved scientific leadership despite their relative lack of riches. Authorial commentary blames international philanthropists for destroying the ambition of Ethiopian people, making them dependent on charity rather than understanding their own wealth.

Far better known than science fiction books in Ethiopia is the Marvel comic series *Black Panther* – or rather the recent movie adaptation of those comics that seems to be familiar to every child, student and technology pundit in the country. The fictional Black Panther is King T'Challa, ruler of the African country of Wakanda. Wakanda is built on a mountain of a wondrous mineral, vibranium, which has many technological uses realized through the advanced knowledge of Wakandan scientists. Dedicated to using the resulting powers only in the cause of peace and prosperity, the Black Panther offers a counter-narrative to contemporary science and technology, in which the potential for benevolent African technocracy can be found in African soil. Although not so well known in Ethiopia until the recent movie production, it has been plausibly suggested that the hero King T'Challa of the original 1966 comic book was inspired by pan-Africanist Marcus Garvey, or even by Emperor Haile Selassie of Ethiopia.

The nationalism of imagined AI

Political leaders are obliged to offer some narrative of progress, but often accompanied by an element of ambivalence regarding the role of technology, since technical change seldom offers equal benefits to all. In the internet era, those benefits that technology does deliver are determined by multinational corporations that often seem beyond the control of national governments. Nevertheless, the current Prime Minister of Ethiopia, Dr Abiy, with a degree in computer science, an MBA and PhD, is clearly aligned with technology. His military career included founding a national Information Security Agency, followed by a government role establishing a centre monitoring science and technology research, and then to a term as Minister of Science and Technology before being elected Prime Minister.

There are two distinctive episodes in which Prime Minister Abiy's name has been drawn into the sociotechnical imaginaries of AI, and these have been mentioned many times by students discussing the significance of AI in Bahir Dar.

Sophia meets Dr Abiy

The first of these episodes is the story of 'Sophia', a humanoid robot who has become an international celebrity. Sophia was created by Hanson Robotics, a Hong Kong-based company that for many years has been a world leader in the manufacture of realistic robotic heads, handmade with a silicon skin, glass eyes and human-like hair. Hanson heads are particularly notable because they incorporate motors to pull the skin in the same manner as human facial muscles, meaning that they can imitate many human expressions. The mechatronic engineering is somewhat challenging, but reasonably conventional – not as complex, for example, as the amazing mechanical automata of the eighteenth century, which were also constructed to imitate human mannerisms. It could be said that the main achievement of Hanson, as with the automata of the eighteenth century, has been in the marketing of these mannequins – not simply as motorized dolls or moving waxworks, but rather representatives of AI. The Sophia model can be mounted on a (stationary) human-shaped body, and its movements synchronized with a speech synthesizer so that it 'talks' with coordinated facial expressions.

International publicity coups for the robot Sophia have included Hanson persuading the Saudi Arabian government that Sophia should be made a citizen of that country. This slightly perverse action (the robot is, after all, little more than a moving doll) gained press attention around the world, with many non-technical commentators on AI now debating the implications of robot citizenship, the implications of Sophia's gender (e.g. Adams 2019), and so on. But in Ethiopia, Sophia has special resonance for two reasons. The first is that one of the founders of a prominent Ethiopian company, iCog Labs in Addis Ababa, provided some of the software components for the robot's expressive movements when he was working for Hanson. The second, and far more significant to Ethiopians, is that Sophia has 'visited' Ethiopia (in boxed parts, some of which went dramatically astray before being reassembled), meeting Prime Minister Abiy, and saying some sentences in Amharic during this meeting. Staff at iCog Labs were also responsible for creating the software for this latter performance.

Press coverage of the event paid great attention to the fact that, while Sophia's 'native' language was English, Amharic was the first language that she had spoken besides English (AfricaNews 5/7/2018; ENA 2/7/2018). Amharic is spoken only in Ethiopia, so despite their pride in its origins and history, Ethiopians seldom see their language represented in any kind of international setting. As a result, the sight of this international celebrity robot speaking their language appeared to have been a profoundly moving event for some - and an event that the technocratic Abiy was very happy to endorse, staging photos of himself engaged in 'conversation' with Sophia, who impressed viewers even more by wearing a traditional Ethiopian dress for the occasion (https://www.africanews.com/2018/07/02/sophia-the-robot-meets-ethiopia-pm-attends-ict-expo). Ethiopian press emphasized that, not only was this a significant occasion for the Amharic language, but the contribution of iCog Labs showed how Ethiopian technologists were in the world class (https://www.rickiebyars.org/z-article-hub-city-live-project).

Building the real Wakanda

The second technological event that has captured the imagination of people in Ethiopia is the launch of a campaign to build the 'real Wakanda' in Ethiopia, inspired by the Black Panther movie, but using the romantic prospect of pan-African technocratic leadership as the starting point for a potentially massive programme of investment in an African technopolis.

Record producer turned entrepreneur Mikal Kamil started his property development career with proposals for the Los Angeles suburb of Compton. Famous outside the USA as the base for hip-hop group NWA, the group's 1988 debut album 'Straight Outta Compton' is considered to have inaugurated the genre of gangsta rap with tracks such as 'Fuck tha Police'. Perhaps unsurprisingly, Compton is today a relatively impoverished neighbourhood, of the kind that might potentially be revived through investment in a 'smart city' centre for arts and entrepreneurship – precisely the proposal that was being promoted by Mikal Kamil as 'HubCity Live!' (PRNewsWire 2017).

The surprising turn of events making that project relevant to this narrative is that Kamil reinvented himself as a technological saviour of Ethiopia, relocating his property development proposal from Compton LA, to a small village outside the city of Bahir Dar (PRNewswire/Hubcity Live-Ethiopia 2018). The smart city concept that he describes as 'Minchu' ('The Source' in Amharic) is no longer a simple arts and entertainment centre, but a new kind of digital/architectural hybrid, transcending the current generations of AI and ubiquitous computing in a new synthesis of African humanity and advanced digital technologies.

A promotional video (HubCity Live! TV 2018) shows the people of Tis Abay crowding the dirt road into the village, holding photocopied signs that read 'Welcome Mikal Kamil/Welcome to Tiss Abay Wakanda/Welcome Mikal Kamil, build the real Wakanda'. According to press releases, a local poet then delivers a powerful sermon:

> When [Prime Minister] Abiy rose from the earth, to deliver his people from immigration and unemployment, Mikal came flying from heaven, like an angel. Tis Abay will become glorious like America. Mikal has arrived so, be happy youth. The American came to live in Tis Abay. Abiy told us to unite with one another. We've always been one and there were no outsiders. We are all united today with America. The Americans came to the waterfall. When they build the technological city, Tis Abay becomes the real Wakanda.

Tis Abay is otherwise famous to Westerners as the village closest to the Blue Nile falls. These are a popular scenic destination for tourists to Bahir Dar, many of whom make the hour-long journey over 30 km of rocky dirt road to see a beautiful waterfall from viewpoints along the hillside sheep trails of the river valley. In the Minchu promotional video, the village of Tis Abay, with its usual donkeys and corrugated iron store fronts out of shot, is replaced by architectural renderings of a dream city - the same renderings, in fact, originally used for Kamil's HubCity Live! promotion from Los Angeles, although now with the 'Jewel of Compton' legend from the earlier promotion edited out.

It seems there are three reasons for the choice of this village as a technologi-cal utopia under the patronage of Prime Minister Abiy. The first is that it has a waterfall, and a waterfall appears as an important ceremonial site in the Black Panther movie. The second is that Tis Abay is (reasonably) near the Amharan regional capital of Bahir Dar, which does have a well-established institute of technology, as well as a reputation among international visitors as an Ethiopian Riviera on the shore of Lake Tana. But the third reason is altogether more fas-cinating, and representative of the ways that the cultural and historical imagin-ation of Ethiopia can become entangled with a technological future.

Students in Bahir Dar suggest that another reason for associating Bahir Dar with Wakanda, beyond the Tis Abay falls, is the rumours of mysterious knowl-edge in the ancient island monasteries of Lake Tana (also a key plot point in the novel *Dertogada*). The lake's status as 'source of the Blue Nile' (the river Abay) is a powerful metaphor encouraging belief in Bahir Dar, and in Ethiopia as the origin of life. A common belief in Ethiopia is that life itself flowed from Ethiopia down the Nile, with fertile soil washed into Egypt via the Abay – the same river described in the book of Genesis as the Gihon, which flowed from the Garden of Eden around the land of Cush.

Mikal Kamil does not hesitate to draw mythological analogies in support of his investment proposal. In an Ethiopian TV interview (Technology and Futur-ism), he describes the Ethiopian people as providing an emotional and spiritual intelligence that can correct the failings of AI, creating an embodied ubiquitous computing that is beyond Western notions of the smart city. He invokes a his-torical Cushite empire that he says lasted (improbably) more than 2 million years during which Ethiopia ruled the known world, and suggests that African-Americans can realize the promise of AfroFuturism only in Ethiopia.

Kamil is a hustler and would-be property entrepreneur, whose views on com-puting and AI draw on mysticism rather than science. Nevertheless, interventions like his really do capture the attention and imagination of Ethiopian people, apparently including Prime Minister Abiy. His sentiment that the magical 'vibra-nium' of the Black Panther movie is within the Ethiopian people themselves, is not so far from an argument for a human-centred AI in Ethiopia. The arguments on mystical grounds do attract opposition, though. Clerics from the local Ortho-dox church suggest that the HubCity proposal may be the work of the devil, and that the technology described as 'inside' Ethiopians might in fact represent plans to tag them with a microchip bearing the Mark of the Beast. If Bahir Dar, or Tis Abay, is indeed the original Eden where a second Paradise is prophesied to come, then Christians worry that Kamil might represent the great Foe of the book of Revelation, come to destroy the monasteries and the site of Eden, not bringing the technological paradise that he claims.

Kamil himself does take some care in addressing these themes, especially in the context of Ethiopia's political sensitivities, which have recently involved vio-lence, church-burning and assassination in the southern Oromo region. Rather than referring directly to the wisdom of Ge'ez that might be preserved in the monasteries of Tana (and which some Ethiopians already fear is being stolen by researchers abroad, including a German research group dedicated to Ge'ez), Kamil invokes the even more ancient empire ruled from Ethiopia, and speaking the Cushite language of the south, rather than the Semitic origins of the church language Ge'ez and Amharic. He avoids suggestions that the African king of the Black Panther comics might have been modelled on the 'Lion of Judah' Haile Selassie, who is seen by some in the South as emblematic of rule by Amharic and Semitic elites. In a political appeal that could easily be shared by Abiy, Kamil says that Ethiopia must combine unity with their sense of duty (and moreover, an AfroFuturist unity, in which African-Americans will join their African brothers and sisters in a new Wakandan empire).

The threat of AI colonization

How attractive would it be for Ethiopians, if the country did become the centre for a new kind of African AI, for this imagined technological advance to be integrated as part of a global digital AfroFuturism?

A prominent feature of the national identity, often mentioned in discussion of national character, is that Ethiopia is the only country on the African continent that has never been colonized (despite residual legacies and tensions over Italian possession of Eritrea and occupation during WWII, and memories of Soviet support for the repressive Marxist-Leninist state of the 1980s). This results in habitual resistance to any analysis of the Ethiopian situation that suggests common features with postcolonial Africa, since Ethiopia never was a colony. It also results in careful concern for autonomy, and sensitivity to the possibility that Ethiopia might be being colonized through less direct means. Historical precedents in the seventeenth and eighteenth centuries included a long period of isolation ruled by Emperor Fasiladas and his successors from the Amharan city of Gondar. Rejection of overtures from the West continued until the nineteenth century reformer Emperor Menelik II enthusiastically adopted modern infrastructure including electricity and plumbing, railways and motor cars, post and banking services, and the telegraph and telephone.

Despite the unquestioned advantages and utility of communications and transport infrastructure to some in the country, the relatively abrupt policy reforms of Menelik left a legacy of some ambivalence regarding the way that an uncolonized country may become reliant on technology, or beholden to those who invent it. That ambivalence continues to be an aspect of current concern about Chinese investment in Ethiopian infrastructure (although this particular concern is also found in other countries of sub-Saharan Africa).

Many countries in the world might have similar concerns about the primarily Anglo-American innovations of the information technology era, and the undoubted power exerted by a very small number of global software corporations that have originated on the east and west coasts of the USA. But most countries have been either colonies or colonizers, meaning that technology is not their primary experience of colonialism. In Ethiopia, which has never been a colony, it is income, infrastructure and technology that present the clearest challenges to continued national autonomy. When it comes to the invisible infrastructure of software and the knowledge economy, Ethiopians are uneasily aware that digital media imposes ways of thinking and arranging their affairs that have not been chosen here (not least, the increasing ubiquity of English as the language of the Internet).

In the case of AI technologies, research students depend heavily on software infrastructure – standard libraries of subroutines and statistical algorithms – that are developed and distributed by a relatively small number of university research groups or (increasingly often) commercial laboratories. There is an uneasy suspicion that the companies distributing such software for free may do so, not only through a spirit of generosity to their academic peers, but perhaps to gain commercial advantage through the de facto establishment and even official adoption of their own technical ideas as future industry standards.

In a country that is sensitive to the colonial dynamics associated with technical infrastructure, reflection on these issues can be uncomfortable for AI researchers. It is not feasible for a relatively small research group to construct its own AI infrastructure – especially not for the intensive and expensive data processing associated with deep learning. The world-leading 'cloud' computation resources of Amazon Web Services are routinely made available for use by students in universities like Cambridge. But a representative of Amazon reports that they are not 'cleared' to operate at all in Ethiopia. A few Ethiopian researchers, with help from grants or private benefactors, are able to purchase the expensive processors necessary to experiment with deep learning algorithms, using open source code libraries obtained from Western labs. But this reinforces the dynamic through which researchers in Ethiopia, using the same tools as the better-funded laboratories that developed those tools, can only follow in the research footsteps of others.

Imagining an Ethiopian AI

Drawing together these threads, of the stories repeated by AI researchers and students, and the practical realities of their historical and political situation, is there a distinctive way of conceiving AI that is both technically and imaginatively distinctive?

Whatever concerns the AI researchers of other countries might have about the redefinition of their questions and evaluation criteria, could AI investors be persuaded to support a use-inspired basic research agenda (Pasteur's

Quadrant) in Ethiopian AI? Although one might expect investors to be comparatively hard-headed, it might not be so difficult to persuade investors as it is to persuade scientists to change their minds – especially if imagination can be recruited.

A rhetorical strategy: from deep learning to broad learning

In seeking ways to stimulate new sociotechnical imaginaries for the AI research community (rather than simply social scientists and critical scholars), it would be necessary to develop rhetorical strategies that are recognizably situated within the technical vocabulary of the AI field, while co-opting conventional terminology to open up new ways of thinking.

One potentially fruitful metaphor for considering new theoretical approaches to AI might draw on the classical distinction in earlier generations of AI research between 'deep' and 'broad' algorithms for problem solving. In algorithms that search for optimal solutions to a problem, two well-known alternatives are 'depth-first search' versus 'breadth-first search'. In depth-first search, the algorithm considers one possible course of action in a branching tree of possibilities, following through the consequences of that choice step-by-step either until the desired outcome is achieved, or else back-tracking to try a different branch if it fails. In the contrasting strategy of breadth-first search, all possible branches are considered and compared, before deciding which one to take first. (Sophisticated algorithms use some combination of these, guided by the structure of the problem or the available data).

We might use these terms to draw attention to the possible research strategies to be followed by AI researchers themselves, when seeking either theoretical breakthroughs or practical applications. At the time of writing, by far the most popular approach to AI research is that of 'deep learning'. Many AI researchers are dedicated to exploring the potential of this neural network technique, which as explained above, is particularly relevant to problems where very large amounts of data and computational resources are available.

The current vogue for deep learning research, and the large proportion of the world's AI researchers digging in this particular hole, might be considered a 'depth-first' strategy in the global research portfolio – if the mine looks productive over here, let's all keep digging until it becomes clear that there is nothing left. Similar gold-rush strategies have been followed in the AI booms of previous decades, usually followed by an 'AI winter', in which disappointed investors and funding agencies withdraw their support from the field.

Is there any alternative to the current 'depth-first' stampede of research into deep learning methods? Might some researchers usefully explore other methods – perhaps a complementary breadth-first investigation that considers 'broad learning' methods rather than deep learning? What might a 'broad learning' algorithm look like (keeping in mind that the metaphorical move from depth-first/breadth-first search to deep-learning does not rest on any technical similarity between depth-first search and deep-learning, but is purely an exercise of rhetorical imagination)? A useful starting place for such a strategy could be to choose a context in which deep learning methods are less appropriate to the problems found there. Perhaps this might be because there are smaller quantities of data available? Or perhaps because there is less computational resource? Perhaps a useful setting in which to explore a new kind of broad learning algoritm might be a country like Ethiopia?

The word 'broad' can be applied in a couple of interesting ways, when considering what 'broad learning' in Ethiopia might imply. The first is breadth in the sense of cultural diversity. Many global AI companies worry that their staff are too homogeneous – too male, too white, too Western. This is not simply because they wish to present themselves as youthful and liberal (although they do wish that), but also because they are aware that a global business relies on a broad range of experience and understanding – especially in the 'knowledge' industries, where a huge company can rapidly disappear if it fails to recognize some change in the zeitgeist. If this is a problem in technology companies, it is also a problem in the international research establishment, and leading international conferences in deep learning such as NeurIPS (Neural Information Processing and Systems) have spawned initiatives including Black in AI (a mainly US-based inclusion group), and the Deep Learning Indaba (a series of summer schools taught in African countries by volunteers from corporate labs and Western universities), to help counter their institutionally narrow cultures.

The second potential application of the phrase 'broad learning', when considering possible strategies for AI in Ethiopia, is to consider the engineering structure of the AI systems themselves. At present, deep learning systems are trained using data acquired from thousands or millions of people – sometimes paid for their work, but more often simply captive customers or unwitting volunteers (Zuboff 2019). But these workers are not invited to reflect on the intention of the resulting system, to deliberate about its decisions, or even to be recognized for their contribution. This has the effect that the kind of 'intelligence' displayed is rather limited. Machine learning researchers prefer to compare the engineering performance of their algorithms by using standard labelled data sets, and do not often ask how the labels were acquired. As argued at the first *Science in the Forest, Science in the Past* workshop (Blackwell 2019a), these practices represent a kind of institutionalized plagiarism, in which un(der)-compensated workers are recruited to carry out intelligent tasks, but after AI systems are built from that data, their owners forget the original work, claiming the intelligence to be solely the work of a machine.

So a 'broad learning' alternative might be to draw on the breadth of human experience, rather than constraining workers to simply provide labels, captions and other simple data to train neural networks in relatively trivial discrimination tasks. Wages in Ethiopia are low, and unemployment is high, meaning that the engineering economics underlying knowledge system design ought to be radically different to the business models of Silicon Valley or Cambridge. A 'broad learning' system could draw in a greater breadth of human understanding through incorporating humans within the economic frame of the system design process, in a way seldom done by conventional AI researchers. Such systems might direct some queries or decisions to human workers, especially in cases where the narrow training of a typical machine learning system would render it unaware of important social or human context.

In fact, although unusual among Western AI researchers, such strategies are not completely unheard of, and certainly not technically infeasible. There is a degree of interest in 'mixed initiative' systems, 'human-centred AI' systems, 'collective intelligence' or 'human-in-the-loop' systems that aim to avoid the worst failings of fully-automated AI through applying a degree of human common-sense. Such approaches could already be used in Ethiopia, although the design principle could also be taken much further. It is quite notable that the approaches most popular in Western AI research tend to appeal to individualistic or market-oriented applications, with theoretical analyses that focus on personal convenience and economic advantage. Collective action in Ethiopian culture is notably different from the individualistic traditions associated with wealthy technology centres, and might well result in different design principles.

Unfortunately, there are a number of reasons why this strategy might not be appealing to AI researchers in Ethiopia. A practical problem is that human 'components' are less reliable and predictable than software algorithms. This would quite probably make such broad-learning AI systems more difficult to design and test than purely software alternatives. Secondly, even if such systems work effectively, how could the quality of the results be compared to the well-cited (and by circular definition world-leading) research coming from the USA or UK? A system that is designed to incorporate a human as one of its components could pass the Turing Test with trivial ease – so much ease that it might make the Test itself seem insultingly trivial. So the use of human 'components' cannot be allowed as a scientific alternative. It would be too easy, and hard science ought to be hard![1]

This quandary relates to the classical problem for science policy that has already been discussed in relation to interdisciplinary innovation and Pasteur's Quadrant. Real insights come, not from a new answer to the same question, but from asking a different question. This is precisely the advantage of doing AI in an Ethiopian context. Perhaps the Turing test was the wrong question? Why do we need AI systems that have no human components? Is it because a human

---

[1] These two sentences are ironic.

component could not be part of a scientifically objective experiment? Or do the commercial investors in AI demand fully-automated solutions so that their software products can be scaled to the global level without the inconvenience of recruitment, personnel problems and so on? How might we argue that a distinctively Ethiopian broad approach to AI is not actually inferior, but superior to the deep-and-narrow alternatives?

Defining and inventing – whose imagination?

This argument has suggested that turning attention to the needs and resources of Ethiopia might result in the necessary conclusion that AI researchers are asking the wrong questions. Changing the question is not always appreciated as the response to a perceived problem, so how might western computer scientists, who consider themselves to have clearly formulated technical questions that define their fields, respond to a suggestion that the field be redefined?

The definition of AI does change, and has continued to change throughout the 70-year history of the field. Until now, the consensus definition has been established in Western countries, and African researchers have followed that lead. Inventing and promoting a different kind of AI does involve establishing a new consensus. But in a field that is so centrally formed by the exercise of imagination, the possibility of a new consensus does seem to be within reach.

In the case of AI, the definition of the technology also incorporates an implicit definition of intelligence itself. An automaton playing chess appeals to a world view in which the game of chess is the pinnacle of intelligent performance (Adam 1998). The *egwugwu* puppets of the Igbo suggest a world in which authority and respect for deliberation is vested in the ancestors that are embodied in those puppets.

So the currently paradigmatic AI systems that have been developed almost entirely in North America can be considered as reflecting new definitions of intelligence in that specific cultural and commercial context. Many of these systems are designed to optimize media attention, marketing and conspicuous consumption, suggesting that such activities have become the paradigms of intelligence in the era of our present digital economy. The consequences of such 'intelligence' for public discourse (whether implemented in mechanical or human form) have become quite clear as these technologies become ubiquitous in Western societies. It is hard to see how the increasingly prevalent commercial logic of surveillance capitalism can be relevant to the bulk of the population, in low income countries that have such low levels of access to ICT.

What of Ethiopia? Is it feasible that, while constructing new digital systems, Ethiopians might adopt different ideals and definitions? The nostalgic and nationalistic kinds of imagination we have described, while departing from conventional research activities in AI, do appeal to alternative imaginaries, where political symbolism and respect for spiritual understanding present an important context of innovation. Pride for tradition and collective responsibility offer a generative principle that is an alternative to the use of ICT for centralized control supported by foreign infrastructure investment. Speculation that Wakanda's fictional 'vibranium' might be a mystical essence inside the Ethiopian people may be sentimental, but in the case of AI, 'intelligence' really is a human property, not a mineral to be discovered through scientific investigation. Any properly scientific enquiry into AI should indeed be looking within ourselves – and certainly not just Western 'selves', but committed to a broad exploration of all human intelligences.

Disclosure statement

No potential conflict of interest was reported by the author(s).

References

Achebe, C. 1958/2010. *Things Fall Apart, Penguin Classics Edition.* London: Penguin. 84–86.
Adam, A. 1998. *Artificial Knowing: Gender and the Thinking Machine.* London: Routledge.
Adams, R. 2019. "Helen A'Loy and Other Tales of Female Automata: a Gendered Reading of the Narratives of Hopes and Fears of Intelligent Machines and Artificial Intelligence." *AI & Society*, doi:10.1007/s00146-019-00918-7.
Blackwell, A. F. 2010. "When Systemizers Meet Empathizers: Universalism and the Prosthetic Imagination." *Interdisciplinary Science Reviews* 35 (3–4): 387–403.

Blackwell, A. F. 2019. "Ethnographic Artificial Intelligence." Paper presented at Science in the Forest and Science in the Past, Needham Research Institute, Cambridge, June 2019.

Blackwell, A. F. 2019a. "Objective Functions: (In)humanity and Inequity in Artificial Intelligence." *HAU: Journal of Ethnographic Theory* 9 (1): 137–146.

Bowman, W. 2015. "Imagining a Modern Rwanda: Sociotechnical Imaginaries, Information Technology, and the Postgenocide State." In *Dreamscapes of Modernity: Sociotechnical Imaginaries and the Fabrication of Power*, edited by S. Jasanoff and S.-H. Kim, 79–102. Chicago: The University of Chicago Press.

Cave, S., C. Craig, K. Dihal, S. Dillon, J. Montgomery, B. Singler, and L. Taylor. 2018. "Portrayals and perceptions of AI and why they matter." (DES5612) https://royalsociety.org/~/media/policy/projects/ai-narratives/AI-narratives-workshop-findings.pdf.

Colburn, T. R. 1999. "Software, Abstraction, and Ontology." *The Monist* 82 (1): 3–19.

Collins, H. 2018. *Artifictional Intelligence: Against Humanity's Surrender to Computers*. Cambridge, UK: Polity Press.

Cox, G., and C. A. McLean. 2013. *Speaking Code: Coding as Aesthetic and Political Expression*. Cambridge, MA: MIT Press.

ENA (Ethiopian news agency). 2 July 2018. "Addis Ababa. Robot Sofia Meets Abiy." https://www.ena.et/en/?p=1677

Gagliardone, I. 2014. "New Media and the Developmental State in Ethiopia." *African Affairs* 113/451: 279–299. doi:10.1093/afraf/adu017.

Gagliardone, I. 2016. *The Politics of Technology in Africa: Communication, Development, and Nation-Building in Ethiopia*. Cambridge, UK: Cambridge University Press.

HubCity Live! TV. 2018. Ethiopia – We Can Build The Real Wakanda – Welcoming Celebration. Accessed Dec 13, 2019. https://www.youtube.com/watch?v=pq4yHf8odk0.

Irani, L. C., and M. Six Silberman. 2013. Turkopticon: Interrupting Worker Invisibility in Amazon Mechanical Turk. In Proceedings of the SIGCHI conference on human factors in computing systems, 611–620.

Jasanoff, S., and S.-H. Kim. 2015. *Dreamscapes of Modernity: Sociotechnical Imaginaries and the Fabrication of Power*. Chicago: University of Chicago Press.

Lloyd, G. E. R. 2007. *Cognitive Variations: Reflections on the Unity and Diversity of the Human Mind*. Oxford: Oxford University Press.

Mumbere, D. 2018. "Sophia the AI Robot Says Ethiopia is a 'Special' Country with 'World Class Talent'." AfricaNews 05/07/2018.

PRNewswire/Hubcity Live!. 2017. "Compton Is Ready For HubCity Live! A Revolutionary New Development Project Destined To Become Urban America's First Technologically Advanced Innovation Hub." Accessed Dec 13, 2019. (https://www.rickiebyars.org/z-article-hub-city-live-project).

PRNewswire/Hubcity Live-Ethiopia. 2018. "Ethiopia, We Can Build The Real Wakanda." Accessed Dec 13, 2019. https://www.prnewswire.com/news-releases/ethiopia-we-can-build-the-real-wakanda-300733455.html.

Schaffer, S. 1999. "Enlightened Automata." In *The Sciences in Enlightened Europe*, edited by William Clark, Jan Golinski, and Simon Schaffer, 126–166. Chicago, IL: University of Chicago Press.

Stokes, Donald E. 1997. *Pasteur's Quadrant – Basic Science and Technological Innovation*. Washington, DC: Brookings Institution Press. 196. ISBN 9780815781776.

Technology and Futurism. "(Ethiopian TV show) uploaded by "HubCity Live! TV". Accessed Dec 13, 2019. https://www.youtube.com/watch?v=T8HULxsW1vs.

Verran, H. 2001. *Science and an African Logic*. Chicago: University of Chicago Press. p. 36.

Wilson, L., and A. F. Blackwell. 2013. "Interdisciplinarity and Innovation." In *Encyclopedia of Creativity, Invention, Innovation and Entrepreneurship*, edited by E. Carayannis, 1097–1105. New York, NY: Springer.

Worku, Y. (trans. Zelalem Nigussie) 2012. *Dertogada*. Addis Ababa: Unity Publishers.

Zuboff, S. 2019. *The Age of Surveillance Capitalism: The Fight for a Human Future at the New Frontier of Power*. London: Profile Books.

Mereological themes in cuneiform worldmaking

Francesca Rochberg

ABSTRACT
Taking as a point of departure Nelson Goodman's idea that it is the frames of reference and the descriptions that refer within those frames that make for ways of worldmaking, this paper considers descriptions of the world attested in Akkadian cuneiform texts as a way to approach the Babylonia *literati*'s notion of 'world.' In this corpus, descriptions of world-parts display mereological themes of complementarity, correspondence, and the relation of counterparts. Meronyms, or names for parts of *the one whole* (we would say 'world,' or 'universe'), and the part-whole relationships they indicate, are of importance to this inquiry into worldmaking.

Introduction

It is a well-known characteristic of Babylonian astronomy that not only did its predictive methods function independently of a physical spatial framework, but its epistemic goals did not include understanding how the world worked as a physical or mechanistic system. If neither a physical structure for the celestial phenomena nor the aim to grasp some structural unity is present in astronomical cuneiform texts, the question of a world conception readily presents itself.

This characteristic of the late Babylonian mathematical astronomy of the second half of the first millennium B.C.E. differentiates the tradition fundamentally from that of its closest historical relative, Hellenistic Greek astronomy, which developed as a direct result of contact with Babylonia. The Greek cinematic form of astronomy is intrinsically tied to a cosmological framework, namely, the spherical geocentric arrangement of the planetary spheres. Babylonian astronomy did not embed its mathematical methods within a cosmos thought to govern the very phenomena the mathematical methods sought to describe. Consequently, we are unable to deduce a cosmological framework from cuneiform astronomical texts.

In order to take account of ideas about a world structure in cuneiform texts, it is necessary to disconnect from the astronomical corpus and focus instead on other sorts of texts, e.g. literary narrative texts, divinatory compendia, divine epithets, names of temples, and royal inscriptions. These afford a picture, if fragmented and partial, about how the world had divine order and organization. These sources do not afford easy access to a unified cosmology either, but in their various descriptions of parts of a world, they offer clues to a way of worldmaking.

In adopting an approach to the notions of world and worldmaking for the cuneiform context, I am borrowing directly from Nelson Goodman's *Ways of Worldmaking* (1978). One of the more interesting, and puzzling, aspects of the cuneiform evidence to be examined here is that rarely is reference made to the world as an integral whole, or a place with explicit boundaries. Not only does this problematize the question of a spatial framework in the Babylonian conception, but also the very question of what ancient Assyrian or Babylonian reference the modern word 'world' really has.

If cuneiform culture's way of worldmaking comes under the rubric 'science in the past,' it underscores the discontinuity between its native tradition and that of the European natural sciences that repudiated and eliminated supernatural phenomena as not belonging to the world science investigates. The idea of the 'external world,' which came to be tied closely to science, has no counterpart in the knowledge tradition of the cuneiform texts. In ancient Assyria and Babylonia a long-standing intellectual tradition aimed to collect and organize knowledge of the phenomenal world. To this tradition belong divinatory texts as well as astronomy, incantation series as well as therapeutic medicine. Where we often go wrong is in equating the phenomenal world of interest to this tradition with phenomena of 'the external world' conceived and explained by modern science, or that the phenomena belonging to the system of reference in cuneiform knowledge should be evaluated in terms of phenomena of an external 'reality' of nature. Instead, the phenomena referred to in cuneiform knowledge comprise an inclusive category of things not limited to natural objects or events. If science is implicated in the basic process of finding order and intelligibility in the phenomenal world of experience and imagination, then the present paper, focusing on the relationship between people and their worlds, their immediate tangible as well as imagined surrounds, concerns itself with science.

As far as the 'science of the forest' is concerned, the relationship between peoples and their worlds, between world and worldmaking, is obviously a fundamental of anthropological interest. Joanna Overing (1990), in an inquiry into the Amazonian shaman as a maker of worlds, embraced Goodman's sense of worldmaking as a productive mode of anthropological understanding. She said,

> the processes of worldmaking … followed in the West and in the jungle are much akin to one another. The scientist, artist, myth teller or historian, and shaman-curer are 'doing much the same thing' in their construction of versions of worlds. However, while the thought processes for constructing worlds are in many ways similar, the facts of which these worlds are made are very different indeed. For instance, in the jungle a world version may comprise angry creator gods and translucent streams of

madness, rather than force fields of energy or atoms and molecules. (Overing 1990, 603)

Following Goodman's claim that, 'our universe, so to speak, consists of these ways [of describing] rather than of a world or of worlds,' I propose to begin with descriptions of the world attested in scholarly cuneiform texts, principally from mythologies of creation, hymnic texts in praise of deities, and royal inscriptions from Neo-Assyrian kings (7th century BCE). By focusing on their ways of describing it becomes possible to discern a complex of themes that point to relationships established between major interrelated parts of the world, such as the regions of the heavens, earth, the netherworld, and the abyss (*Apsû*).

Prominent among these interrelated themes are complementarity, correspondence, and the relation of counterparts. Meronyms, that is, names for parts of *the one whole* (we would say 'world,' or 'universe'), and the part-whole relationships they indicate, are of importance to this inquiry into world-making that is based on words, terms and descriptions and the things/world parts to which they refer. In looking at meronyms, or names for parts of the whole, it is interesting to note that the expected holonym meaning 'world' in fact does not emerge. Instead we see a variety of words for 'totality, all, or entirety (of parts).'

In all cognizance of the variation in descriptions and representations, with especially wide variation diachronically, this paper pursues a thematic analysis from Akkadian meronyms (names for parts) and meronoms (the parts themselves) in order to understand how mereological world-parts, such as they are described, were set in relation to a whole. It further raises the question of how, or in what way, out of the many terms for 'all of x (parts),' there was a notion of the whole, as that is an essential part of meronomy, which deals with part-whole relations in contexts where there is a set of parts forming a whole.

Worldmaking in cuneiform

The present investigation of worldmaking in the cuneiform culture of the first millennium B.C.E. proceeds from descriptions found in literary and scholarly texts. In Goodman's words (1978, 2–3): 'If I ask about the world, you can offer to tell me how it is under one or more frames of reference; but if I insist that you tell me how it is apart from all frames, what can you say? We are confined to ways of describing whatever is described.' By virtue of such descriptions we establish a relationship between frames of reference and systems of description. How the first millennium Babylonian *literati* understood the structure and composition of their world is described in mythological narratives (see Horowitz 1998; Katz 2003; Lambert 2013). The description of the world, or its parts, found in these sources construct a world version true to its frame of reference.

The literary sources for world-description are paramount, but what is not so obvious is how far those conceptions of the world and its component parts extend into non-literary sources. The question, then, of the relationship of world-versions is problematic. Mythological narratives and god-lists of cuneiform culture have been approached as representative of the rudimentary beginnings of the history of cosmology, from the standpoint that historical systems can be assessed by comparing them to the 'right rendering' of the world of modern science. In the case of ancient Mesopotamia, the world-version attested in the stories of gods and the Creation was once viewed as going nowhere, as in the following statement by W.G. Lambert (1975, 49):

> In many ways it is a disappointment that the civilisation which produced so much in the collection of data and in the abstract sciences did not develop its cosmological ideas during the 3,000 years of its existence, and the reason for this can be sought. Two main factors were at work. One was the old, prehistoric mythological thinking, which, despite the fictional element in its products, was properly based on sound observation of the universe. This could have provided the basis for a scientific cosmology if conditions had been right. The other was the personification of natural forces into anthropomorphic gods and goddesses, and this led away from the realities of nature into theological fantasy. Unfortunately the growth of civilisation resulted in a diminution of the first of these two factors and an increase in the second.

Lambert found that cosmological ideas were stunted for thousands of years because of 'old, prehistoric mythological thinking' and 'personification of natural forces.' This statement, a relic of scientism, separates science ('realities of nature') from religion ('theological fantasy'), two modern categories without counterpart in cuneiform antiquity.

The imagery of world-parts such as the watery *Apsû* or 'Abyss' (home to the god Ea) and the imagined *Ešarra* (home to the god Enlil) was not of interest to mathematical (or observational) astronomical texts, which as mentioned before, operated independently of a spatial cosmology. References to world-parts are found in texts that relate directly to the gods, such as in a late Assyrian prayer to the god Marduk, where the major constituent parts of the world are named as entities that should 'witness' that god's deeds:

> May all the gods and goddesses, Anu, [Enlil], the constellations (*lumāšu*), the Abyss, the netherworld ... behold the deeds of the lord of the gods, Marduk. (*līmurū epšet bēl ilī Marduk ... lumāši apsû daninnu* Livingstone 1989, 2:36–37).

Babylonian astronomical ephemerides and procedures appeal to none of the places such as are named in the exhortation just quoted, although they do have cause to mention the *lumāšu*-constellations, which in astronomical contexts denote the special category of 'counting stars' (MUL.ŠID.MEŠ = *lumāšu*) for marking the paths of the moon and planets in nightly observation. Different frames of reference produce different systems of description.

When it comes to the question of world-structure (Lambert 1975, 2013; Horowitz 1998; Geller 1999; Katz 2003), the key text is the poem *Enūma Eliš* 'When Above.' The text was recovered in 1849 when Austen Henry Layard found the clay tablets containing *Enūma Eliš* in the ruins of Nineveh. In 1876 George Smith brought out an English edition, and in 1890 Peter Jensen's *Die Kosmologie der Babylonier* set the text within the wider scope of cosmology. At the start of the twentieth century, the British Museum published L.W. King's cuneiform copies of the text in volume 13 of *Cuneiform Texts from Babylonian Tablets in the British Museum* (1901). King's edition and translation, titled *The Seven Tablets of Creation, or the Babylonian and Assyrian Legends concerning the creation of the world and of mankind*, appeared in rapid succession in 1902. The authoritative edition appeared in Lambert 2013. The poem, 'When Above,' is perhaps the finest example of cuneiform literature and scribal scholarship, not only in terms of its high poetic style, but also and notably its scholarly mastery of cuneiform orthographies and hermeneutics. It belongs at the centre of any discussion of the Babylonian conception of world structures according to first millennium scribes.

When we look at the narrative of creation from *Enūma Eliš* Tablet IV, we see that the story is principally a moral one, and yet there are the world-structural components explained by the narrative. Marduk, the hero, slays the goddess Tiamat, whose body will become the heavens. In killing Tiamat, Marduk vanquishes evil, treachery, and improper divine rule. *Enūma Eliš* accuses Tiamat of having contempt for mercy, for appointing her own spouse, Qingu, to 'the rank of Anuship' (the highest divine station) though he had no right to that office. She is further accused of giving Qingu the Tablet of Destinies,[1] though he had no right to hold it. She is charged with stirring up trouble and perpetuating evil against the gods. In the battle Marduk is aided by an incantation for undoing evil (Akkadian *tû*, Sumerian TU$_6$) as he approaches his foe: (*Enūma Eliš* IV 60-62) 'He set his face toward the raging Tiamat. In his lips he held a spell (*tû*) and in his hand he held a plant to counteract (Tiamat's spell?).' Tiamat as well, when she found she was up against the redoubtable Marduk, first turned into a raving maniac (*Enūma Eliš* IV 88 *mahhûtiš ītemi ušanni ṭēnša* 'like one insane she lost her mind') and then repeatedly recited an incantation to protect herself (*Enūma Eliš* IV 91). The moral victory was complete when Marduk rounded up Tiamat's divine helpers, smashed their weapons, threw them all in a net and confined them to prison (*Enūma Eliš* IV 111-114). The culminating triumph was when the hero Marduk wrested the

---

[1]The mythological Tablet of Destinies is defined this way by Andrew George (1986, 138): 'The theological exposition of the function and nature of the Tablet of Destinies … makes use of the traditional terminology associated with the tablet. It is … the means by which supreme power is exercised: the power invested in the rightful keeper of the Tablet of Destinies is that of chief of the destinydecreeing gods … , which amounts in principle to kingship of the gods. The poet of the Anzu Epic makes this clear when he describes Anzu's usurpation of supreme kingship by theft of the tablet, and the concomitant upsetting of the cosmic status quo.'

Tablet of Destinies away from Qingu, whose possession of it had been an infraction against divine propriety (*Enūma Eliš* IV 121).

When after the Babylonian *Chaoskampf* the creation of the world is finally mentioned in *Enūma Eliš* IV 135-136, the god rests and surveys 'the corpse' of Tiamat, which is also referred to as a *kūbu* 'lump' (of flesh, usually in reference to a stillborn or premature fetus, or a monstrosity of some kind). Having then fashioned the heavens from half of her tightly stretched body, and keeping her waters from escaping by installing watchmen, he proceeded to make heaven as a counterpart measured to the size of the subterranean watery *Apsû*. Finally, he rectified the order of divine propriety in setting up the shrines where the high gods were to have their residences (*Enūma Eliš* IV 141-146). The major parts of the world that are created in order to house these deities and set the world aright are the heavens (home to Anu), Earth (as Marduk's new residence), the *Apsû* (home to Ea), also called Ešgalla, and Ešarra (literally 'House of the All') for Enlil.

Enūma Eliš provides an authoritative picture of the creation of the world, but variations in world-structural imagery are also attested (Rochberg 2020), just as in the Classical and Greco-Roman worlds more than one cosmology took shape (Pythagorean, Milesian, Platonic, Atomist, Aristotelian, Stoic, Ptolemaic). Before the various philosophical traditions pursued inquiries into the nature, structure, and material constituents of the cosmos, notional underpinnings of the word *kosmos* (κόσμος), as Jaan Puhvel's etymological inquiries revealed, are derivable from Homer as well as from Cretan inscriptions (Puhvel 1976, 154–57). These contexts, he said (1976, 154), 'point to a notion of ordering, arraying, arranging, and structuring discrete units or parts into a whole which is 'proper' in either practical, moral, or esthetic ways.' The semantics of order, ritual procedures, and cultic correctness are similarly reflected in the Akkadian word *parṣu*, used by the scribes to translate Sumerian GARZA, a word frequently associated with Sumerian ME, whose uses go back into third millennium sources. ME's semantic range covers concrete to abstract things, from divine attributes and offices, to rites and norms, but also including drawings or emblems, presumably for the abstractions ME can reference (Hrůša 2015, 35–38).[2] Akkadian *parṣu* can also be used to mean emblems, symbols, and insignia of abstract things such as divine authority, the power of kingship or other office (CAD *parṣu* mng. 4).

The significance of this complex of words and meanings is that before *kosmos* took on the meaning 'world, universe' with ontological implications, there was a more active sense of the order and harmony established by correct ritual procedure and the rightful carrying out of divine (or royal)

---

[2]Hrůša (2015, 35, note 32) points out that the Oxford ETCSL translates ME as 'essence,' but he argues that this, as well as the translation 'being,' is a pseudo-Platonic importation into the semantics of ME. The abstractions that ME could connote belong to the realm of things that establish norms, such as 'rites' and 'offices,' not ontological considerations of being.

offices. Certainly in cuneiform literature, the order and stability of the world was expressed in terms of its functioning rather than its material form.

System of world description: meronyms and meronomies

In the fields of semantics and pragmatics, according to Alan Cruse, a meronomy is a 'type of branching hierarchy' of the 'part-whole type' (Cruse 1986, 157). The prototype for meronomies is the division into parts of the human body, where the whole in that case is the 'body' occupying the apex of the hierarchy and underneath are parts such as 'head,' 'arm,' etc. These, then, have their own parts that branch underneath, e.g. 'nose,' 'hand,' and so on. Meronymy, on the other hand, considers, as Cruse put it (1986, 159), 'the lexical items used to designate parts and wholes, and the semantic relations between them.' Expressions for part-whole relations might, then, point to ways that the ancient cuneiform scribes considered questions such as, what it means for a thing to have parts, for something to exist in the world, how existing things and their parts are classified, and the mereological sums, or wholes, that result from such thinking about what we now call parthood.

In one important respect the cuneiform evidence makes such analyses extremely difficult. The first and most formidable stumbling block, in my view, is that the very idea of 'matter' is not within the sights of the scribes. The kinds of levels of existence we assume, such as the physical and the non-physical, not to mention the idea of matter itself, or the idea of a combination of formal and material properties (e.g. Aristotelian hylomorphism), are a world away.[3] Thus the ontological implication for ideas about material things, and what makes them divisible into parts, or composed of such parts, will be quite different from what generally comes of more standard Western philosophical mereologies, even of the 'Neo-Aristotelian' kind (Koslicki 2007).

It could rightly be asked, as a result, what kind of ontological system can be derived from such limited structural terminology as is available in cuneiform texts. These terms reflect ideas about how world-parts are arranged and structured, but not how the 'nature of matter' or of 'material things' determines that they should be composed of parts at all, or in what way. We find no talk of essences or essential properties because there was no interest in matter or material existence such as would give priority to physical things or their ability to cause other physical things to occur or change in the material world.

The focus on meronymy/meronomy, or the names of parts and the parts themselves, makes possible a comparison, or counterpoint, with more familiar approaches to understanding the cosmos from the Greek world. Even in

---

[3]Indeed, the word we translate as 'form' in Akkadian, *alandimmû*, is a Sumerian loanword from ALAN 'form, statue' and DÍM 'to create, fashion.' But the 'form' in Akkadian refers to the outward appearance of something, even the human face and body, as in the physiognomic omen series *Šumma alandimmû* 'If the form.' It has no semantic use relative to the internal or intrinsic properties determining the shape or form of a thing.

otherwise quite divergent cosmological schemes, the part-whole relation is clear in the notion of 'cosmos' that underpins the particular Greek way of worldmaking exemplified in the one all-encompassing world that persisted into the later Christian and Ptolemaic version.[4] A case in point is Aristotle's definition, or rather three definitions (*De caelo* 1.9 278b9–21), of *ouranos* from the work *On the Heavens*. These are (1) the substance of the outermost revolution of the universe of the world and the seat of divinity; (2) the body containing the moon, sun, and some of the stars; (3) the body enclosed by the outermost revolution, i.e. the entire world, that is, the universe, *ouranos*. (Leggatt 1995, 88–89) The passage concludes with the remark that 'the whole world which is enclosed by the outermost circumference must of necessity be composed of the whole sum of natural perceptible body, for the reason that there is not, nor ever could be, any body outside the heaven' (Aristotle, *On the Heavens*, Bk A, Ch. IX, 278b, 10-25).

Another locus for the idea of a unified cosmos is found in Stoicism. Brooke Holmes (2019, 245) notes, 'The conspicuous fact of cosmic sympathy compels us in turn, the Stoics insist, to accept the reality of less obvious things, such as the unity of the body of the cosmos.'[5] She also observed (2019, 240) that the Stoics'

> theorization of the cosmic organism enacts a scenario in which the idea of Nature as a totalized whole is taken to its absolute limit … In Stoic sympathy, we find a pious commitment to the unity, completeness, and perfection of the world together with a tarrying with its parts and its particulars.

Such 'totalizing' and unification, or ideas concerning the absence of anything outside the world, are foreign to Assyro-Babylonian scholarly traditions.

A number of Akkadian words mean 'all,' 'totality,' or 'entirety,' but in each case the word is constructed with another noun in the genitive for a sense of 'all of' some set of things included within it or counted as parts of it. For example, a word meaning 'all' or 'totality' (*kiššatu*) is used precisely in the sense of 'all of x,' or 'the totality of x.' In these constructions *kiššatu* is found together with some other noun in the genitive, as in 'all (of the) lands' (*kiššat mātāte*), 'the entire sky' (*kiššat šamê*), 'all mankind' (*kiššat nišī*), or 'all the gods' (*kiššat ilī*).[6] Alone the term *kiššatu* connoted 'the entire inhabited world,' as in the well-known royal title 'king of the universe' (*šar kiššati*), but there *kiššatu* certainly has the sense of all known inhabited places, not 'all' in the sense of all that exists. Divine epithets for Marduk and Nabû use the phrase 'all heaven and earth/

[4]Other Greek and Roman cosmological systems, such as that of the Atomists, and particularly the Epicureans, had the idea of multiple worlds, even an infinity of worlds (Warren 2004). The possibility of the Pre-Socratic Anaximander and Anaximenes speculating on an infinity of *kosmoi* is also entertained, though the evidence is limited (Warren 2004, 354 note 2).
[5]Holmes quotes Sextus Empiricus as saying that from such observations as the phases of the moon coordinating with phenomena on earth, 'it is obvious that the cosmos is a unified body,' in Sextus Empiricus, *Against the Professors* 9.79; apud Holmes (2019), 245 note 17.
[6]See the CAD s.v. *kiššatu* for references.

netherworld' (*kiššat šamê u erseti*) as a designation of worldwide power, but the meaning is conveyed by the construction of 'all of' with the merism of 'heaven and earth/netherworld.'[7]

A notable aspect of the term *kiššatu* is that it can be written with the logogram ŠÁR,[8] which, as the word for the number thirty-six hundred, is further used to mean an indefinite large number, and thus in a certain sense, 'totality.'[9] Sometimes its meaning is 'countless,'[10] and so by extension 'all.' The latter, however, is limited to the expression *adi šāri* to convey a sense of 'for all time' or 'everywhere,' mostly in bilingual literary contexts. These expressions convey a sense of temporal as well as spatial 'totality,' as in the passage that refers to the radiance (*melammû*) of kingship that overwhelms 'everywhere' (*adi šāri*, see CAD s.v. *šār* mng 2).

In a note on the origin of the Greek term *saros*, Otto Neugebauer (1969, 141) said that Sumerian ŠÁR meant 'universe or the like' as well as the number thirty-six hundred. He explained the crooked path by which the number 3600 came to give its (Sumerian) name to the eclipse relation 223 synodic months = 242 anomalistic months, which is the basis for prediction of eclipses every approximately 18 years, or 6585 and 1/3 days as a standard value. From the point of view of usage, and available attestations, the Akkadian loan *šār*, however, does not fill the gap in terminology for 'universe.' On the other hand, the Sumerian word ŠÁR, represented in its earliest form graphically as a circle, may indeed have connoted something of an encompassing 'all.' As mentioned before, the name Ešarra (É.ŠÁR.RA), literally 'house of the 3600,' or 'house of the all' becomes a cosmic location as the abode of the Sumerian creator god, Enlil.

Another word (*kullatu*), similarly, means 'all,' or 'totality,' and is frequently constructed with a genitive of parts that can be counted, e.g. 'all the gods' (*kullat ilī*), 'all peoples' (*kullat nišī*), or 'all cities' (*kullat māhāzi*). In extended meaning there is 'the whole of all wisdom' (*kullat nagbi nēmeqi*), and, perhaps most general in semantic force is the construction with the idiom 'everything, anything,' literally 'whatever its name' (*mimma šumšu*), to mean 'all of everything.' Esarhaddon says of the god Ea, lord of wisdom, that he was 'the one who fashioned everything, whatever its name' (*pātiq kullat mimma šumšu*, Leichty 2011, 104:4). In this case, 'whatever its name' (*mimma šumšu*) is an abstract way of designating 'all parts,' and so is similar to the previous examples where *kullat* + gods, lands, or peoples is the way to express 'all of x.' Finally, from a prayer to Marduk by Assurbanipal, a construction with another lexeme meaning 'totality' (*hammatu*, from hamāmu 'to gather') gives us

---

[7]See below for discussion of the dual meaning of *ersetu* as both 'earth' and 'netherworld.'

[8]*šá-ár* ŠÁR = *kiš-ša-tum* Idu II 70, CAD s.v. *kiššatu* lexical section.

[9]Going back to Sumerian, see the *Electronic Pennsylvania Sumerian Dictionary*, s.v. *šar* [3600].

[10]Especially in the phrase *šār bēri*, literally '3600 double-hours,' used to mean 'a countless number of x,' see CAD s.v. *šār* meaning 1b.

'possessing the whole of all wisdom, the totality of all strength' (*hammata kullat nēmeqi gamir emūqi* [*gašrāti*(?)], see Livingstone 1989, 7:4).

The substantive *kalu* similarly means 'whole, entirety, all,' and is sometimes constructed following a noun referring to things that can be counted, as in 'signs, all of them' (*ittātim kalašina*), or 'prediction from birds, all (kinds) of them' (*purussû iṣṣūrū kalašunu*). The extended meaning of the word *napharu* 'all, whole, totality,' stems from its basic meaning 'sum, total (of things counted or added),' which occurs in the context of a total tally of items in a list, or the sums of numbers.[11] It is sometimes found in absolute use,[12] e.g. 'the totality of people, all of them' ([*te*]*nēšetu ša naphari kališunu*), or to describe rulership over (*ana malikūti*) everything, or one who is appointed to be in charge (*pāqid*) of everything. More often it is found in construction with *kullatu*, as in 'the totality of everything' (*kullat naphari*), also preceding a genitive such as 'all of Assyria' (*naphar māt Aššur*), 'all of Chaldea' (*naphar māt Kaldi*), or 'the totality of gods' (*naphar ilī*), and 'all the black headed (people) as many as exist' (*naphar ṣalmat qaqqadi mala bašû*). Finally, a poetic word *nagbu* means 'totality' or 'all,' in the same way as its synonymns *kullatu*, *napharu*, and *kiššatu*.[13] Just as in the case of its synonyms, *nagbu* refers to all 'of something,' constructed with what is being totaled in the genitive, such as 'all the Aramaeans' (*nagab Ahlamê*), 'all his enemies' (*nagab zāʾerīšu*), 'all the swamps' (*nagab berate*), 'all living beings' (*nagab napišti*). Already mentioned is also 'the whole of all wisdom' (*kullat nagbi nēmeqi*).[14]

Somewhat different is the use of the term 'regions' (*kibrātu*), often constructed in the expression 'the four quarters' (*kibrāt erbetti*), which then represented 'the entire inhabited world.'[15] In one of its usages, the phrase 'the four regions' (*kibrāt erbetti*) has a political rather than a cosmological connotation, used in royal titulary and in inscriptions describing either hostilities in or domination over 'the four regions.' The notion that the four regions represent the entire world reflects a royal ideology in which divine appointment to kingship is in fact to rule over the entire world, as in the claim to be the one 'whom Assur appointed to rule the entire world' (*ša Aššur … ana muʾirrūt kibrāt arbaʾi šumšu … išquru*).[16]

The last terms to consider here, *gimru* and *gimrētu*, from *gamāru*, whose basic meaning ranges from 'to finish,' 'encompass,' 'complete,'[17] present a somewhat different sense of totality as compared against the previously discussed terminology. The key passage is in the scene in *Enūma Eliš* where the

---

[11]CAD s.v. mng. 1.
[12]CAD s.v. mng 2 usage a.
[13]It is included in the lexical synonym, or thematic, word list Erimhuš (Tablet 5 line 44), and is given as a synonym of *napharu* in the commentary to the Akkadian Theodicy (BWL 74, comm. To line 57).
[14]All references are given in CAD s.v. *nagbu* B.
[15]CAD, s.v. *kibrātu*.
[16]CAD s.v. *kibrātu*, usage a 1' b'.
[17]Also 'to bring to an end,' as in 'annihilate.' See CAD s.v.

assembly of the gods give Marduk kingship 'over the entirety of the whole of everything' (*kiššat kal gimrēti*), which he then demonstrates by annihilating and then recreating a star (or constellation, using the term *lumāšu*, *Enūma Eliš* 1 IV 14). Another such passage is from the *Akītu*, or New Year's, ritual: 'You (Marduk) survey everything with your eyes, you watch over all omens/ oracles/decrees/extispicies (*têrtu*) through the decrees (*têrtu*) you give' (*ina īnēku tabarri gimrētu* [*ina*] HAR.BAD.MEŠ-*ka tahâṭu* HAR.BAD.MEŠ[18]). Another reference to *gimrētu*, or *gimrātu* is in a lexical synonym list where *gi-im-ra-tum* = MIN (=[*pu-uh-ru*]),[19] and *puhru* stems from *pahāru* meaning to gather things together. In another passage for *puhru*, the plurality of things making up 'the whole' is clear: 'Anu created all (meaning the collection of all) the heavens, Anu created (the collection of) all the earths' (*Anu puhur samê Anu puhur erṣeti ... ibtani*, see Köcher BAM 538 ii 52).

The foregoing examples underscore the difference between the concept of a totality expressed by 'all *xs*,' and the concept of a whole that functions within a lexical part-whole hierarchy, or meronomy (Cruse 1986, 157). Among the terms for 'all of x (parts)' discussed above, none offers a true holonym for the whole of what the parts comprise. Nor is it possible to construct a branching hierarchy from the world parts as belonging to the whole of which each is an autonomous part. Given that Cruse (1986, 171) said, 'there are no meronomies of unnamed wholes,' it is, therefore, difficult to know how to analyse the various world parts (heaven(s), earth, netherworld, *Apsû*) as meronoms (parts) or the words for them as meronyms (names of parts) of something that has no word or name (holonym) for itself. In the absence of a straightforward meron-omy, then, let us consider the *themata* that relate the parts to each other, and, perhaps then, indirectly, indicate a notion of the whole.

Themata: complementarity, counterparts, and correspondence

It is clear that the passages containing the many Akkadian terms for 'all' (*kullatu, kiššatu, kalu, napharu, nagbu, gimru, gimrētu* and *puhru*) do not focus these descriptions on the internal structure or material substance or even overall shape of the world as a whole, but rather as the sum of its many parts or constituent things. It is my sense that a meronomic branching hierar-chy cannot be reconstructed. More common are references to parts of the world in various non-hierarchical binary relations to one other, most basic of which are the regions of heaven(s) and earth(s). The idea that certain parts of the world, such as 'above' and 'below,' complement one another, forming by virtue of their relation one whole, is an essential ingredient of the cuneiform conception of the world.

---

[18]RAcc 130:20.
[19]CT 18 21 Rm.354:7.

The use of merisms, comprised of terms for contrasting or complementary parts, e.g. 'heaven and earth,' to express what is conceived of, presumably, as a whole, is evident in the cuneiform sources as it is commonly in other languages (Lloyd 1966; Wasserman 2003). Parts that complement one another, perfecting and making a whole from those parts, are often at the same time viewed as counterparts of one another. These two *themata* can therefore be considered together, as indeed the complementary parts often constitute counterparts.

The complementarity of two principal parts seems to be the oldest and most basic of conceptions of 'world.' The idea is preserved in the terms AN.TA.MEŠ (Akkadian *elâtu*) 'the upper region(s)' and KI.TA.MEŠ (Akkadian *šaplâtu*) 'the lower region(s),' and the idea of upper and lower halves of the world reflects an original mythology of creation by the separation of AN 'above' from KI 'below' (Lambert 2013, 169–171). The complementary parts heaven and earth, taken as a pair in Sumerian as AN.KI 'heaven and earth' (as well as the Sumerian AN.ŠÁR and KI. ŠÁR or 'all heaven' and 'all earth') and in Akkadian as *šamû u erṣetu* 'heaven and earth (or netherworld)' remain throughout cuneiform cultural history as the most basic structural duality for the world as a whole.[20] Wasserman (2003, 82) analyses the common merism of 'heaven and earth' as referring to a 'vertical totality of space,' which is a wholly different concept from one of 'universe.'[21]

The notion of 'above and below' is, moreover, basic to the way the world is made through celestial divination, that signs above are equivalent to signs below. They are counterparts, as expressed in a singular text known as the 'Diviner's Manual,' which puts it as follows: 'the signs on earth as well as of the sky bear (lit. 'wear,' *našû*) omens[22] for us; heaven and earth alike bring us omens; they are not separate from one another; heaven and earth are interconnected. A sign that is evil in heaven is evil on earth; that which is evil on earth is evil in heaven.'[23] This describes a world wherein the counterparts of good and evil are manifested in the two complementary realms of the world, construed as above and below.[24]

---

[20]Steinkeller (2005, 18–23) discusses the two halves of the world as 'hemispheres,' making a spherical universe. This, I think, is to over-read the sources as something more like that of the classical Greek conception since Eudoxus of a spherical world totality. Apart from that detail, his reconstruction of the cosmological background to the extispicy ritual (*ikrib mušītim* 'nightly ritual') is cogent.

[21]Jeremy Black (1998, 152) also observed the use of names in Sumerian for the extremes of the parts of the world to express *per merismum* the entirety of the space between them.

[22]This line uses two different words for 'sign' or 'omen,' *ittu* and *ṣaddu*, the latter signifying especially signs produced by deities, including the celestial deities manifested as planets (moon, Jupiter). See the CAD s.v. *ṣaddu*.

[23]Oppenheim (1974, 200:38–42).

[24]This complementary conception is not the equivalent of the particular macrocosm/microcosm idea that wherein the human being is a reflection of the larger whole, namely of the universe of the celestial spheres contained by the sphere of the fixed stars. This cosmic imaginary provides the basis for Greco-Roman astrology, and alchemy as well. Melothesia texts from Late Babylonia certainly foreshadow this notion of the reflection of the heavens in the individual. I would take these as examples of the theme of correspondence rather than as evidence of a spatial spherical conception with humanity at the center. For the melothesia texts, see Geller (2014), also Wee (2015).

The complementary status of heaven and earth, however, is complicated by the fact that the 'below' of this equation (the lexical equation KI = *erṣetu*) denotes either earth or the netherworld, depending on context. A few sources confirm the identification of the *erṣetu* as the place where netherworld deities resided; thus *erṣetu* means 'netherworld' in such contexts. In other contexts *erṣetu* clearly means the earth as dwelling place of the living (Horowitz 1998, 273–4, and see the passage from *Atra-Hasīs*, 'I cannot [set my feet on] the earth of Enlil,' Lambert, Millard, and Civil 1999(1969), 91:48). Noegel (2017) suggests that in the Bible (Gen 1:1), as well as in *Enūma Eliš*, the merism should be understood as heaven and netherworld, not heaven and earth. This, he argues, is the result of the use of the term *erṣetu* for both earth and netherworld, and that the structure that makes sense of the terminology is a five-level organization of the world into heavens (*šamû*) – starry sky (*šamû*) – earth (*erṣetu*) – fresh water abyss (*Apsû*) – netherworld (*erṣetu*), wherein the words *šamû* and *erṣetu* each have two references. (Noegel 2017, 124)

The complementarity of heaven and earth/netherworld (*erṣetu*) is found in a motif often repeated in temple hymns and royal inscriptions concerning the building of ziggurats and temples, whose tops were so high as to 'rival heaven' and whose foundations were so deep as to reach into the netherworld (*erṣetu*). As an example, 'he (Sargon) established its (the temple Eanna's) foundation in the depth of the netherworld like a mountain.' (*temmenšu ina irat kigalla ušaršid šadū'aiš* YOS 1 38 I 40) It was also a way of expressing great magnitude, as in the description of a mountain encountered by Sargon II (721-705 BCE) on campaign, 'whose summit above leans against the heavens, and whose base, below, is firmly rooted in the netherworld' (*ša eliš rēšāša šamāmi endāma šaplānu šuršūša šuršudu qereb aralli* TCL 3, p.6: 19.) The motif was also applied to mythological mountains in the Gilgamesh Epic (George 1999, 71 IX 38–41) that reached up to the base of heaven and down to the netherworld. It is also found in the Erra Epic, where the sacred tree, whose wood was suitable for constructing a statue of the supreme god Marduk, is similarly described as having roots in the netherworld and a crown that touches the highest heaven (Foster 1996, 765).

The conception of the ultimate heights and depths of a sacred tree, a mountain, or a temple had a variant where earth/netherworld is replaced by the depths of the abyss itself, the *Apsû*. The metaphor is traceable to the early third millennium BCE in an archaic hymn: 'Great, true temple, reaching the sky, temple, great crown, reaching the sky, temple, rainbow, reaching the sky, temple, whose platform(?) is suspended from the midst of the sky, whose foundation fills the *Apsû*.' (Biggs 1971, 201). The *Apsû* was the place of the watery abyss, not the salt sea, rather the sweet fresh waters of the springs (*nagbu*[25]), as

---

[25]This *nagbu* is a different lexeme from its homophonous *nagbu* B, discussed above. For *nagbu* 'spring' or 'source,' see CAD s.v. *nagbu* A.

we know from Sumerian and Akkadian mythology, divine epithets, and hymnic literature. The unceasing, seemingly eternal, fresh water springs and source of the rivers no doubt accounts for the inclusion of the *Apsû* as one of the cosmic regions. Thus in the Hymn to Šamaš, about the man who does good deeds, it is said '[his] descendants will go on forever like the eternal spring waters (*nagbu*).'[26]

The innermost part of the fresh water abyss was beyond the reach of human perception. No one had plumbed the depths of the *Apsû* nor seen its interior (Horowitz 1998, 317). The depth of the *Apsû* was an unknown region, as described in a broken line that compares something (in the broken beginning of the line) to the unknowable inner regions of the *Apsû*, saying 'that, like the inside of the distant *Apsû*, not even a god can find (it).'[27] Although a part of the world largely inaccessible to human perception, the *Apsû*'s sweet waters in rivers and springs that bubble up from underground were places where the *Apsû* reached the earth's surface.

Apsû's watery depths were associated with the creation, abode, and kingdom of the god Enki (Sumerian)/Ea (Akkadian), so closely associated with him that Enki/Ea's son, Marduk, was known as 'first-born son of the *Apsû*.' Because of Enki/Ea's association with wisdom, magic, and incantations, the *Apsû* was the fount of wisdom and source of the secret knowledge of incantations. The temple of Enki in the oldest Sumerian city of Eridu was called the E-Abzu 'House of the Abyss.' In *Enūma Eliš*, Marduk's temple Esagil in Babylon was said to be the counterpart (*mihirtu*) of *Apsû* (*Enūma Eliš* VI 62).[28] As we have already seen, *Enūma Eliš* IV 142 places Ešarra, the dwelling place of Enlil, as a counterpart (*tamšīlu*) of Ešgallu, the 'great shrine' or *Apsû*. And the ziqqurat foundation, Etemenanki (its name itself testimony to the idea of the complementarity of above and below) is said to be a counterpart (GABA.RI) of Esharra in an inscription of Nabopolassar (VAB IV 62 iii 19; cited Lambert 2013, 200).

One of the great sources for the notion of counterparts is *Enūma Eliš*, when Marduk demonstrates his authority in a wholesale reorganization of the world and the settling of gods in it. Equally as important, the poem accounts for the location of the city of Babylon, and site of Marduk's own temple Esagila there, as a new cosmic centre. Marduk's residence in the Temple Esagila effectively placed Babylon at the centre of the world. Marduk's 'house,' the Esagil, is said to be 'the equivalent' or 'counterpart' to the great abyss, the *Apsû*, where Ea dwelled: *Enūma Eliš* VI 62 'They raised the peak of Esagil, a replica of the *Apsû*' using the term *mihirtu* 'equivalent, counterpart.' The importance of

---

[26]Lambert BWL 132:121 *kīma mê nagbi dārî zēra[šu] dā[rî]*.

[27]See CAD s.v. (*w*)*atû* lexical section for the passage [...] sù.ud.du.gin$_x$ dingir na.me.nu.mu.pàd.da.e.ne : [...] *ša kīma qereb apsî rūqi ilu mamman la uttû*, translated there as 'who, like the innermost of distant *Apsû* no god can discern.'

[28]Also Esagil as the gaba-ri ap-se-(e) in VAB VII 300 10; cited by Lambert (2013, 200).

the idea of measured counterparts, equivalents, is also clear in *Enūma Eliš* IV 141–46 'He (Marduk) crossed the heavens, surveyed the sky, and he adjusted them as the equivalent of the *Apsû*, Nudimmud's abode. Bel measured the shape of the *Apsû* and set up Esarra, the counterpart to Esgalla. In Ešgalla, (in) Ešarra which he had built, and (in) the heavens, he settled Anu, Enlil, and Ea in their shrines.'

A further indication that the idea of counterparts played a thematic role in a cuneiform world description is the use of the term *maṭṭalātu* from a verb (*naṭālu*) meaning 'to look at, face, or point toward.' The word is relatively rare, occurring only in first millennium scholarly or literary contexts. The clearest usage is no doubt the one found as the incipit of Tablet 16 of the liver omen series (*Barûtu* 'the art of inspection [of the ominous entrails]'), which is: *šumma amūtu maṭṭalāt šamê* 'If the liver is an image/counterpart of heaven.' (CT 20 1:31) Another attestation of *maṭṭalātu* occurs in the context of the account of rebuilding of the temple Esagil by Esarhaddon. In his royal inscription, in a deliberate reference to the Babylonian *Enūma Eliš*, Esarhaddon says the temple Esagil is the counterpart of the *Apsû* (*maṭṭalāt apsî*) and the equivalent, or likeness (*tamšīlu*) of Ešarra (Leichty, 2011, p. 198 Esarhaddon 104 iii 41b-iv 1, and p.206 Esarhaddon 105 iv 37b-v 15). The passage is worth quoting in its entirety:

> [In] a favorable month, on a propitious day, I laid its (Esagil's) foundation platform over its previous foundations (and) in exact accordance with its earlier plan I did not diminish (it) by one cubit nor increase (it) by half a cubit. I built and completed Esagil, the palace of the gods, an image (*maṭlāt*) of the Apsû, a replica (*tamšīl*) of Ešarra, a likeness (*mehret*) of the abode of the god Ea, (and) a replica of (the square of the constellation) The Field; I had (Esagil) ingen[iously] built (and) I laid out (its) square (*mithartu*).[29]

In the context of world regions, the thematic notion of the counterpart refers to temples, mountains, trees, and the liver, and it connects the divine, or heavenly, world with something important for human life. When the temple of Ea, the Ešarra, is set up as the replica or counterpart (*mihirtu*) of the Field constellation, or the Square of Pegasus, the celestial quadrant marked by the stars is established as the model in heaven for the earthly shape and foundation for the Ešarra temple in Babylon. The best indication that the meaning of *mithurtu* is 'opposition,' in the sense of complementary counterpart, is in the writing of the Sumerian word that is translated as *mithurtu*, namely HA.MUN, meaning 'harmony (of opposites),' which is spelled with the Sumeriogram NAGA.-NAGA (where the second NAGA is written upside down).

Continuing the binary nature of *themata* attested in many world descriptions from cuneiform scribal scholarship is that of correspondence. The theme of

---

[29]The reading could also be *mithurtu*, which has the meaning 'opposition,' as in *lišān mithurti* 'contrasting tongues or conflicting opinions, in which case the work might be read *mithartu* 'square.' See the discussion section under *mithurtu* in CAD s.v.

complementarity between celestial and terrestrial applies in divinatory interpretation, where evil above is evil below, and so on, but correspondence forms a different relationship constructed of signs and their consequents,[30] the protases and apodoses of omen texts. Every omen then, in making a relationship between phenomena from one domain to those in another, instantiates the *thema* of correspondence, so this is perhaps the best attested of all *themata* in cuneiform knowledge.

Typical examples of correspondence can be given where deities are set in corresponding relationships with phenomena, for example with the four winds: South (corresponds to) Ea, the father of the gods; East (corresponds to) Enlil, the lord of all; North (corresponds to) Ninlil, lord of phantoms; West (corresponds to) Anu, father of heaven. (George 1992, 152–3) The parts of the liver correspond to specific deities, and the months to zodiacal signs, as in this example from a late text from Uruk:

> The Presence (corresponds to) Enlil, the month Nisan, [the Hired Man (Aries)]. ...
> The Path (corresponds to) Shamash, the month Ajaru, The Bull of Heaven (Taurus) ... The Pleasing Word (corresponds to) Nusku, the month Simanu, True Shepherd of Anu (Orion) ... [...] The Strength (corresponds to) Ninurta, the month Du'uzu, the Crab (Cancer) ... and so on. (Koch-Westenholz 2000, 24–25)

Correspondence can in fact be copiously documented, continuing well into late medical astrology between zodiacal signs and parts of the human body (Heessel 2005; Geller 2014; Wee 2015), not to mention the correspondences given in the microzodiac texts between dodecatemoria (the twelfth parts of a zodiacal sign), temples, stones, and woods (Weidner 1967; Heessel 2005). The relation of correspondence between particulars, or parts, of whatever system is in question – lexicography, omens, astrology, medicine – is too prevalent for us not to take this into account in how we understand part-whole relationships in cuneiform culture. Also, and no less significant, is that through the correspondences established in divinatory texts what is of interest to human beings is brought into relation with the world around, be it the heavens (*Enūma Anu Enlil*), the behavior of animals (*Šumma ālu*), the plants (medical texts, late astrological medicine), or stones (medical and ritual texts, late astrological medicine).

Conclusion

The *themata* characterizing the descriptions of world parts, namely complementarity, counterparts and correspondences, are non-hierarchical binary relationships. The fact that there is no Akkadian counterpart to our notion of 'world,' but rather identification of a totality is always as the sum of parts, as in 'all of x (things),' raises the question of whether these elements of world

---

[30] I use the term consequent for the event brought into relation to a phenomenon in the conditional 'If P, then Q' formulations of Babylonian omens. A better way to understand the conditional relationship expressed by omens is as 'If P is the case, Q is the case,' without a causal connection.

description are effected through meronomy. Meronomies are typically based on part-whole relationships, but the nature of those relationships are not narrowly defined, either with respect to spatial, structural, material or functional relations of various parts to the whole. What they do have are distinct ideas of those part-whole relations, necessitating a designated name (holonym) for the whole in relation to the parts. In the passages from cuneiform texts in which world-parts are described, the interest appears to be less about the part-whole relationship and more about the part-part relationship, specifically about binary relations in the form of counterparts, complements, and correspondences.

On no level do we find examples of branching hierarchies of parts in their relation to an overall whole. Nor can we find evidence that such wholes were ontologically prior to its parts. Totalities of parts constituted wholes, as lands were sums of all inhabited cities. Another kind of totality is found in the sum of many things, e.g. all wisdom or 'the entirety of all of everything,' both of which connote sums rather than unitary wholes.

The exploration of mereological themes in cuneiform worldmaking shows a focus on non-hierarchical binary relationships and the tendency to disregard questions of substance and composition in favour of correspondence, complementarity, and correlation. In view of Goodman's statement (1984, 14, emphasis in the original) that 'there is no one correct way of describing or perceiving "*the world*,"' it is important to be able to account for the particularity of the cuneiform ideas through their own way of worldmaking, and also to consider that different traditions within the cuneiform corpus refer to different 'worlds.' We are then able to leave the idea of a physical external world out of that picture, to focus instead on what constituted the ancients' right rendering, or right description, of the world in its own context and within its own frame of reference. Goodman's view (1978, 3–4) that '"the world" depends upon rightness,' advocates for a multiplicity of 'rightness of renderings of all sorts.' If the anthropologist of historical sciences can establish from the evidence 'what makes a version right and a world well-built' (Goodman 1984, 29) the 'science of the past' can open onto more diverse and multiple kinds of knowledge both within and across cultures.

Acknowledgements

My sincere thanks go to the readers of this paper, G.E.R. Lloyd, Lorraine Daston, and Nicholas Jardine, who all saved me from errors of omission and commission and helped me to clarify innumerable points. All references to Akkadian passages use the standard abbreviations for cuneiform publications found in the *Chicago Assyrian Dictionary* (CAD).

Disclosure statement

No potential conflict of interest was reported by the author(s).

References

Biggs, R. D. 1971. "An Archaic Sumerian Version of the Kesh Temple Hymn from Tell Abu Ṣalabikh." *ZA* 61: 193–207.

Black, Jeremy. 1998. *Reading Sumerian Poetry*. London: Athlone Press.

Cruse, D. A. 1986. *Lexical Semantics*. Cambridge: Cambridge University Press.

Foster, Benjamin R. 1996. *Before the Muses, Vol. II*. Bethesda, MD: CDL Press.

Geller, Mark. 1999. "The Landscape of the 'Netherworld'." In *Landscapes, Frontiers, Territories and Horizons in the Ancient Near East, vol. III. CRRAI (Compte rendu, Rencontre Assyriologique Internationale) 44*, edited by L. Milano, S. de Martino, F. M. Fales, and G. B. Lanfranchi, 41–49. Padua: Sargon.

Geller, Mark. 2014. *Melothesia in Babylonia: Medicine, Magic, and Astrology in the Ancient Near East*. Boston, Berlin, Munich: De Gruyter.

George, Andrew R. 1986. "Sennacherib and the Tablet of Destinies." *'Iraq (1977)* 48: 133–146.

George, Andrew R. 1992. *Babylonian Topographical Texts*. Orientalia Lovaniensia Analecta 40. Leuven: Peeters.

George, Andrew R. 1999. *The Epic of Gilgamesh: The Babylonian Epic Poem and Other Texts in Akkadian and Sumerian*. London: Allen Lane/Penguin Press.

Goodman, Nelson. 1978. *Ways of Worldmaking*. Indianapolis, IN: Hackett Publishing Company.

Goodman, Nelson. 1984. *Of Mind and Other Matters*. Cambridge, MA: Harvard University Press.

Heessel, Nils. 2005. "Stein, Pflanze und Holz. Ein neuer Text zure 'medizinschen Astrologie." *OrNS* 74: 1–22.

Holmes, Brooke. 2019. "On Stoic Sympathy: Cosmobiology and the Life of Nature." In *Antiquities Beyond Humanism*, edited by E. Bianchi, S. Brill, and B. Holmes, 239–270. Oxford: Oxford University Press.

Horowitz, Wayne. 1998. *Mesopotamian Cosmic Geography*. Winona Lake, IN: Eisenbrauns.

Hrůša, I. 2015. *Ancient Mesopotamian Religion*. Münster: Ugarit-Verlag.

Katz, Dina. 2003. *The Image of the Netherworld in the Sumerian Sources*. Bethesda, MD: CDL Press.

Koch-Westenholz, Ulla. 2000. *Babylonian Liver Omens: The Chapters Manzazu, Padanu and Pan Takalti of the Babylonian Extispicy Series Mainly from Assurbanipal's Library*. Carsten Niebuhr Institute Publications, 25. Copenhagen: Museum Tusculanum Press.

Koslicki, Kathrin. 2007. "Towards a Neo–Aristotelian Mereology." *Dialectica* 61: 127–159.

Lambert, W. G. 1975. "The Cosmology of Sumer and Babylon." In *Ancient Cosmologies*, edited by Carmen Blacker and Michael Loewe, 42–65. London: George Allen and Unwin Ltd.

Lambert, W. G. 2013. *Babylonian Creation Myths*. Winona Lake, IN: Eisenbrauns.

Lambert, W. G., A. R. Millard, and Miguel Civil. 1999(1969). *Atra-Hasīs: The Babylonian Story of the Flood, with the Sumerian Flood Story by M. Civil*. Oxford: Oxford University Press. and Winona Lake, IN: Eisenbrauns.

Leggatt, Stuart. 1995. *Aristotle: On the Heavens I-II*. Warminster: Aris & Phillips.

Leichty, Erle. 2011. *The Royal Inscriptions of Esarhaddon, King of Assyria (680–669 BC)*. Royal Inscriptions of the Neo–Assyrian Period 4. Winona Lake: Eisenbrauns.

Livingstone, Alisdair. 1989. *Assyrian Court Poetry and Literary Miscellanea*. State Archives of Assyria 3. Helsinki: Helsinki University Press.

Lloyd, G. E. R. 1966. *Polarity and Analogy: Two Types of Argumentation in Early Greek Thought*. Cambridge: Cambridge University Press.

Neugebauer, Otto. 1969. *The Exact Sciences in Antiquity*. 2nd ed. New York: Dover.

Noegel, Scott. B. 2017. "God of Heaven and Sheol: The 'Unearthing' of Creation." *Hebrew Studies* 58: 119–144.

Oppenheim, A. L. 1974. "A Babylonian Diviner's Manual." *Journal of Near Eastern Studies* 33: 197–220.

Overing, Joanna. 1990. "The Shaman as a Maker of Worlds: Nelson Goodman in the Amazon." *Man* 25: 602–619.

Puhvel, Jaan. 1976. "The Origins of Greek Kosmos and Latin Mundus." *The American Journal of Philology* 97: 154–167.

Rochberg, Francesca. 2020. "Mesopotamian Cosmology." In *Blackwell's Companion to the Ancient Near East*, edited by Daniel Snell, 305–320. Oxford: Blackwell Publishing.

Steinkeller, Piotr. 2005. "Of Stars and Men: The Conceptual and Mythological Setup of Babylonian Extispicy." In *Biblical and Oriental Essays in Memory of William L. Morn*, edited by A. Gianto, 11–47. Biblica et Orientalia. Rome: Istituto Biblico.

Warren, James. 2004. "Ancient Atomists on the Plurality of Worlds." *The Classical Quarterly* 54: 354–365.

Wasserman, Nathan. 2003. *Style and Form in Old-Babylonian Literary Texts*. Cuneiform Monographs 27. Leiden: Brill/Styx.

Wee, John. 2015. "Discovery of the Zodiac Man in Cuneiform." *JCS* 67: 217–233.

Weidner, E.F. 1967. *Gerstirn-Darstellungen auf babylonischen Tontafeln*. Sitzungsberichte, Österreichische Akademie der Wissenschaften 254. Graz: Böhlau in Kommission.

Monteverdi's unruly women and their Amazonian sisters[1]

Stephen Hugh-Jones

ABSTRACT

Lévi-Strauss argues that Amazonian mythology reveals a moral philosophy concerned with control of entries and exits to the tubes and apertures of the body. But which body? Following clues from attitudes to female singers in renaissance Italy, this essay suggests that this Amerindian body is not the one we take for granted but rather one very similar to the body that figures in the writings of Galen and other classical authors. This insight sheds new light on Amerindian mythology, where tubes and apertures are a dominant theme, and on ritual where music and tobacco smoke lend substance and fertility to life-giving breath and control of body orifices is emphasised. Such ideas, rooted in a common-sense anatomy and physiology, are also familiar in the modern West but have been submerged in Elias' civilizing process.

Introduction

In his 1964–1965 lectures on Amerindian mythology, Claude Lévi-Strauss outlined 'a moral philosophy preoccupied with certain immoderate uses of the digestive tube – positively or negatively, above or below' (1984, 111 *my trans*). This philosophy underlies Amerindian mythology about the origins of pottery. Lévi-Strauss elaborates further on this philosophy in his book *The Jealous Potter* (1988). The book explores an analogy between pots and the physical and moral state of the women who make them. Thus, as pots should be closed and retentive, myths about the origins of pottery typically concern women who are jealous, i.e. retentive in another mode. Furthermore, menstruation is frequently believed to be incompatible with pot-making, for pots should be watertight whilst menstruation suggests leakage.

To explore this theme further, Lévi-Strauss shows us how Amerindian myths about pottery use various birds and animals to encode the open or closed state of the body's orifices. Here, nightjars, with their gaping mouths, and noisy howler-monkeys, who frequently defaecate on hunters pursuing them on the

[1]Based on the Société d'Ethnologie Eugène Fleischmann lecture 'Les Femmes indisciplinées de Monteverdi et leurs sœurs amazoniennes' (Musée Quai Branly, 12 November 2019). I wish to thank Michel Adam, Carlos Fausto and Geoffrey Lloyd, for their helpful information, comments and suggestions. Any errors are mine not theirs.

ground below, both connote incontinence and openness. In contrast, the sloth connotes closure and retention. To appreciate why this is so, one must first know that sloths rarely make any noise, that their only cry is a high-pitched scream, and that they have fastidious toilet habits. Sloths rarely defaecate and, when they do so, they come all the way down from the trees where they live in order to leave piles of hard, round pellets on the ground below.

Now, as we shall see, sloths and howler monkeys and the theme of open or closed body apertures figure prominently in the story of a deity known as Jurupari that is told by the Tukanoan and Arawakan-speaking peoples living in the upper Rio Negro region of NW Amazonia.[2] The name 'Jurupari' comes from Nheengatu, a lingua franca once widely used throughout NW Amazonia. Arawakan speakers refer to him as Kuwai whilst he has a number of different names in the many different Tukanoan languages spoken in the region. The Jurupari story concerns a deity who is burned on a fire. From his ashes a palm tree springs up that is then cut into sections to make flutes and trumpets, instruments identified with Jurupari's dismembered bones and used in secret men's initiation rites. Women are obliged to hear the sound of these instruments but they must never see them. However, in the story, the women succeed in stealing the flutes from the men.

Curiously enough, neither the Jurupari story nor musical instruments figure anywhere in Lévi-Strauss' book. This is especially surprising given the importance that Lévi-Strauss gives to music elsewhere in his work. The reason for this omission lies in the fact that, for Lévi-Strauss, the mythology of NW Amazonia bears the hallmarks of a self-reflective, savant philosophical tradition associated with the complex civilizations of the middle Amazon region (1973, 271–272). This mythology thus belongs to a different genre from the simpler, more spontaneous popular mythology that forms the subject of his *Mythologiques*. Thus, instead of NW Amazonian mythology, *The Jealous Potter* has as its principle focus the myths of the Jivaro and other peoples of the eastern sub-Andean piedmont region, a mythology that bears a striking resemblance to that of the indigenous peoples of Southern California. One of the concerns of Lévi-Strauss' book is to demonstrate how, in their mythology, Amerindian peoples who live in widely separated parts of North and South America nonetheless deal with the same philosophical issues in very similar ways – so much so that, for Lévi-Strauss, the Amerindian mythology of the two Americas is one.

One result of this focus on the Jivaro and their neighbours is that, for Lévi-Strauss, the artefact that best exemplifies his digestive-tube philosophy is the

---

[2]My own research involves the Barasana, Bará, Eduria/Taiwano, Makuna, and Tatuyo, groups speaking eastern Tukanoan languages who live along the equator in south-eastern Colombia just across the frontier with Brazil. For the purposes of this paper and to avoid ponderous repetition of 'Northwest Amazonia', 'Tukanoan and Arawakan', and 'Tukanoan', I use the shorter 'Tukano' to refer to all speakers of eastern-Tukanoan languages and treat the mythology of Tukanoan and Arawakan-speaking groups as forming a single common tradition.

blowgun, a weapon whose South American distribution coincides with that of the philosophy in question. The blowgun and digestive tube also have a technological connection: breath from the mouth first blows a poisoned dart up the tube of the blowgun to kill an animal. The meat that results is then eaten through the mouth, travels down the digestive tube and is later expelled through the anus as excrement (Lévi-Strauss 1987, 87).

In this chapter, I want to re-consider Amazonian body-tube philosophy in the light of the Jurupari story. I will suggest that, in the figure of Jurupari, we can find much of Lévi-Strauss' argument condensed succinctly into a single story and that it is the flute, rather than the blowgun, that best exemplifies this philosophy. Indeed, with respect to the peoples of NW Amazonia, one might summarize the philosophy in question by saying simply that 'the body is a flute'.

In connection with this, and despite a wealth of brilliant insight, the body that figures in Lévi-strauss' book remains, in essence, the modern, western body that is familiar to us. Traces of this modern body are also evident in much of the extensive literature devoted to the body in the ethnography of Amazonia. In this chapter I want to suggest that we can take Lévi-Strauss' insights further when we realize that the anatomy and physiology at issue in mythology from NW Amazonia is that of a rather different body and that the range of substances that are considered to flow in and out of the apertures of this body extends beyond what, to us, might seem to be the most obvious candidates. Aside from air, saliva, tears, blood, semen, urine, and excrement, these substances include not only the hair that sprouts from the crown and fontanelle of the head but also the vision that appears to stream from the eyes to illuminate the world around,[3] and the sounds that spew from the mouth to penetrate the ears of the listener. These are not new ideas. In fact, they are embedded in our own cultural past. And this is where *Monteverdi's Unruly Women* take their place on stage.

In her book of the same title (2004), musicologist and historian of science Bonnie Gordon sets out to demonstrate how the anatomical and physiological theories of Galen, Aristotle, Hippocrates and other classical authors that were still in vogue in sixteenth and seventeenth-century Italy can shed new light on how female singers and their songs were understood by the people of that time. I want to suggest that this classically informed, late Renaissance body, at once physically and intellectually closer to our own scientific and cultural roots and also the subject of explicit philosophical reflection, may help shake us from our modernist assumptions and allow us to perceive the outer form and inner workings of the different body that is apparent in the mythology and ritual of NW Amazonia.

---

[3]The bright yellow, reflective eyes of the jaguar provide a striking example here.

In his *Mythologiques*, Lévi-Strauss is never shy of drawing audacious parallels between the mythology of the Americas and the historical ethnography of Europe. In juxtaposing the unruly, flute-stealing women of Amazonia with their unruly singing sisters of the Italian Renaissance, I feel that I am in good company. Let us begin with Galen.

Galen's body

For the purposes of this chapter, I shall use Galen's name and ideas as a catch-all that stands for a number of different medical and philosophical writings that all tend in roughly the same direction. For brevity, I shall also reduce some complex and often disputed ideas to three simple principles. I must stress that am no classicist and do not have time to enter into a detailed discussion.[4]

The first principle is that of *horizontal analogy*. For Galen and others, the male genitals were an everted version of their inverted, female counterparts, the vagina corresponding to the penis, the labia corresponding to the foreskin, the womb corresponding to the scrotum, and the ovaries to the testes. The everted, more complete genitals of men were believed to result from the greater heat, dryness and vitality of men's bodies, a sign of men's superior status and greater moral worth and testimony to their place in a hierarchical social and cosmic order. In the colder, wetter bodies of women, the vagina remained inverted and internal. In his book *Making Sex*, Laqueur (1990) summarizes these ideas as a one-sex, two-gender model of the human body and argues that the development of our contemporary two-sex, two-gender model coincides with the transition from the mediaeval era to modernity. However, King (2013) disputes this claim, providing solid evidence that the one-sex and two-sex models are already both evident in the medical writings of classical antiquity.

The second principle is that of a *vertical analogy* between the upper and lower halves of the body. Here the mouth and throat are paired with the vulva and vagina as twin passages leading to the stomach-womb understood as a single organ. Consistent with all this, it was believed that 'first intercourse deepens a woman's voice and enlarges her neck which responds in sympathy with the stretching of her lower neck' (Armstrong and Hanson 1986, 99). It was also believed that one could test if a woman was pregnant by making her squat over a pot of garlic and testing the smell of her breath.

The third principle is that of the *fungibility of body substances*. Here the different body fluids involved in digestion, respiration and conception, were considered to be transformations of each other with wellbeing depending on a balance between wet and dry and hot and cold substances, the basis of the humoral system. Blood and semen were seen as equivalent and complementary

---

[4]My discussion relies on secondary sources including Armstrong and Hanson (1986), Gordon (2004), King (2013), LaMay (2007), Laqueur (1990), Preuss (1997) and Wikipedia (2019).

fluids and, as the vehicles of *pneuma* that circulated in the body as breath, soul, consciousness and vitality, both substances were considered necessary for conception. However, men's hotter breath meant that only male semen had the capacity to generate life. Consistent with these ideas, Soranus of Ephesus also taught that, because female singers exhausted precious blood in their songs, they tended not to menstruate and were often infertile.

Finally, for classical medicine, hair was also understood to be a derivative product of blood or blood in another form. As we will see below, hair has a special significance in the Tukano version of body-tube philosophy that I am concerned with here. In Tukano mythology, hair figures as another substance that flows from holes in the body.

According to Gordon (2004), in early modern Italy, the bodies of women were imagined as chronically leaky vessels. Female singers were thought of as taking this natural incontinence to a transgressive, erotic extreme. With breath as an already hot and thus 'male' substance, hot, saliva-laden breath emerging from the wide-open, mobile mouth and beating throat of a female singer was imagined as a substance in motion, something akin to both blood and semen, that penetrated and titillated the ear of the male listener and gave him pleasure. With such thinking in mind, those who wrote songs for these singers would choose the sense of the words and the articulation of the vowels so they played upon the correspondence between the singer's mobile mouth and her genitals.[5] For reasons such as these, female singers were considered to be sexually forward, transgressive and unruly beings who usurped the singing role normally associated with men. No wonder then that parents were unwilling for their daughters to become singers. Should these daughters persist in going on stage, they were made to undergo regular checks on their virginity.

Let us turn now to Amazonia and to the story of Jurupari where we will meet another band of unruly women, the flute-playing sisters of Monteverdi's female singers. In NW Amazonia, principles similar to those outlined above are played out in a very different cultural context so we should not expect to find exact parallels. What matters is that, taken overall, the parallels are sufficiently close for us to use the expression of these principles in one context as clues to understanding their expression in another. Bearing this caveat in mind, I think that the Jurupari story and associated rites suggest that something akin to the early European ideas I have outlined above also applies in Tukano mythology and ritual.

The story of Jurupari

In this brief chapter I can only provide a brief, composite summary of what is, in fact, a long story that comes in many different versions.[6] My summary aims

---

[5]See also LaMay (2007).
[6]For an excellent discussion and comprehensive bibliography of relevant sources on the Jurupari story see Vianna (2017).

to show the elaboration of the body-tube theme, one that is alluded to in the very name Jurupari. Straddelli (1929, 498) glosses the name as 'that which closes our mouth', *iruru* being 'mouth' and *pari* is a large screen of strips of paxiuba palm-wood woven together with vine and used to block the passage of fish by closing-off the mouths of streams or lakes. Such screens are also used to make *cacuri* fish traps and the enclosures in which young men and women are confined during puberty rites. The story that follows would suggest that the 'Jurupari' who eats (or anally ingests) initiates and retains them in his belly incarnates this very enclosure. An alternative gloss for his name would thus be 'enclosure mouth'.

The story divides into three episodes: the birth of Jurupari; the death of Jurupari; and the theft of the flutes that come from a palm that springs from his ashes and are identified with his bones.

The birth of Jurupari

At the beginning of time, a deity created ancestral humans by blowing cigar smoke onto powdered coca leaves in a womb-like gourd (Figure 1). Later this deity spied on a male deity who was concealing cigar-ash in a pot – other versions talk of coca or tapioca in a gourd. She then stole and ate the ash/coca/tapioca and became pregnant. However, in spite of her open, prying eyes and a greedy mouth, our deity also lacked the genitals that would allow her to give birth. To help her, the men then used a cigar holder to provide her with the missing parts (Figure 2). Much later in the story, this same deity

Figure 1. *(Left)* Yebá Buró creates a new being from cigar smoke (Drawing by Luiz Lana. Lana and Lana 1995, 67). *(Right)* Tukano man smoking a cigar in a cigar-holder (photo Koch-Grünberg 1967, 281).

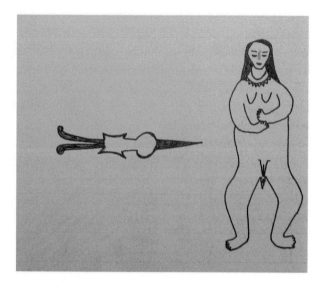

Figure 2. *(Left)* Cigar-holder (photo Koch-Grünberg 1967, 282). *(Right)* Cigar holder as genitals (Drawing by Luiz Lana. Lana and Lana 1995, 234).

becomes a sexually voracious ogress with scorpions and biting ants in her vagina.

When the baby Juruparí was born he had no mouth so the men made one for him, first as a vertical slit[7] and then the proper horizontal version. As soon as he had a mouth, this baby began to give out loud cries and terrible sounds. He also had an insatiable thirst. The men stole the baby from his mother and gave him to a sloth foster-mother, the antithesis of the baby's excessively open mouth. However, the baby Juruparí quickly sucked his sloth foster-mother dry before ripping off her breast.

Juruparí grew very fast. A constant stream of hair and sound flowed from the holes or pores of his body along with both terrifying noises and sweet music.[8] Juruparí is also described as having the body of a rainbow boa or other giant constricting snake. As a large, brightly-coloured and strikingly-patterned boid snake that is full of all kinds of music, Juruparí is the direct counterpart of the Anaconda who figures widely in Amerindian mythology, sometimes as the source of all designs and sometimes as the source of the colours of birds.[9] As we shall see, sounds, colours, birds, and designs come together again in the figure of Kaapi discussed below.

Juruparí is also sometimes portrayed as having the feet, hands and hairy body of a sloth, a cannibal mouth with jaguar fangs and the voice of a

---

[7]Evoking the slang term 'vertical smile'.
[8]One prototype for this extraordinary hirsute being who becomes a palm tree may well be the shaggy fibres, (used to make ropes and brushes) that cover the trunk of the piaçaba palm. 'Piaçaba' translates as 'hair emerging from the heart of a tree' (Meira 2017, 129).
[9]See, for eg. Barcelos Neto (2016), Lévi-Strauss (1970), 302-319; Oliveira (2015); Van Velthem (2003).

howler monkey. In this guise he is also the source and master of poisons and sickness. At other times he appears in more benign form as a handsome dancer with a feather crown on his head and with sounds flowing from his body. Luiz Lana, a Desana artist, renders these sounds in visual form as plume-like curlicues reminiscent of the speech-scrolls of Mesoamerican iconography (Lana and Lana 1995, 235) (Figures 3 and 4).

We can see then that Jurupari's body condenses within it a set of extreme, opposed elements such as music and hair, sound and colour, closure and openness, and life-giving musical instruments and death-dealing poisons. It also contains within it all the animals and trees in the forest and all the elements of the cosmos. The one exception here is fire. Announcing that 'only fire can kill me', Jurupari goes up into the sky.

The death of Jurupari

(To follow this part of the story, the reader must first know that it recaps the stages of initiation.) At initiation, young boys are shown the Jurupari flutes for the first time. After this, they are secluded in a special enclosure, made from either woven palm leaves or a screen of woven palm-wood splints (*pari*), where they must fast and have no contact with fire or women. This symbolic death is followed by their re-emergence from seclusion in a symbolic rebirth.[10]

Later, Jurupari returned to earth and revealed himself to some young boys. As they had now seen Jurupari for the first time, they were thus in the status of initiates and required to fast. Jurupari, now in the form of a howler monkey, reappears to tempt the boys by throwing down uacú (Monopterix uacu) fruit from a tree. The seeds of these large, bean-like fruit are only edible if they have first been roasted. Unable to resist, the boys roast and eat the uacú beans.

Enraged by the smell of fire and roasting and by the boys' disobedience, Jurupari reverted to his dangerous, open, cannibal state. As his voice sounded like thunder, saliva and tears poured from his mouth and eyes in the form of heavy rain. Seeking shelter from the storm, the boys rushed into a 'cave', either Jurupari's open mouth or, in other versions, his open anus (Figure 5). With the boys secluded inside his belly, Jurupari then returned to the sky. Soon the boys' bodies of the boys began to rot, a transformation that parallels the transformation of initiates in seclusion.

Jurupari refuses to return the boys to their parents so the men resolve to kill him. They tempt him back to earth by inviting him a dance-feast where they promise to serve his favourite beer. Unable to resist, Jurupari returns to earth and gets drunk at the feast. As the drunken Jurupari vomits the boy's bones into an enclosure already prepared for his arrival, they are reborn as initiates

---

[10]For a full description and analysis of Jurupari rituals see Hugh-Jones (1979).

Figure 3. Kuwai-Juruparí (*Kuwai-ka-Wamundana*) as cannibal sloth, owner of sickness and sorcery. (Drawing by Thiago Aguilar 2010. Wright 2018, 23).

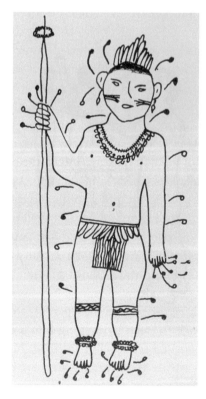

Figure 4 . Juruparí as dancer with music coming from his body (Drawing by Luiz Lana. Lana and Lana 1995, 235).

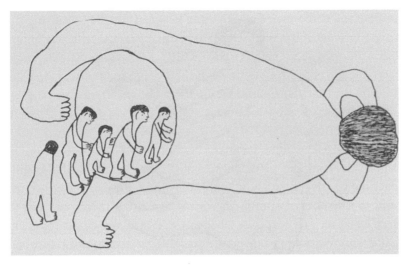

Figure 5. Juruparí devours the disobedient boys by causing them to enter his anus. (Drawing by Luiz Lana. Lana and Lana 1995, 236).

who emerge from seclusion and come back to life. The men then kill the cannibal Juruparí by pushing him into a fire. A palm tree springs from his ashes and grows up to the sky, taking Juruparí's soul with it. The men cut this palm into sections which they distribute to the different human groups as the musical instruments that they now use in their initiation rites, the transformed bones and dismembered body of Juruparí that are kept hidden underwater in the river when not in use.

The theft of the flutes

Much later, a prying young woman overhears her father ordering her brother to rise early to bathe and vomit in the river, something that young men are expected to do each morning. The lazy boy stays asleep but his sister goes to the river where she finds one of the forbidden flutes. Putting the instrument between her legs, she tries in vain to play it. A big-mouthed fish then comes along and shows her how. She gets together with the other women and together they steal all the men's flutes and feather ornaments, and thus recuperating, in another form, the baby Juruparí who was stolen from them in the beginning. By stealing the flute-baby, the women also cause an inversion of the social order. The women now take over the men's ritual activities and begin to take the initiative in sexual relations whilst the men not only do the women's heavy work of growing manioc and processing the tubers but also begin to menstruate. Today, when initiates see the Juruparí instruments for the first time, people say it is as if they too are menstruating.

The 'proper' order is only restored when the men succeed in getting back their flutes. Some versions of the story have it that, to punish the women for their disobedience, the men forced a flute between their legs, causing them to menstruate instead of the men. Other versions have it that it was the women who hid a flute in their bodies. This is why women menstruate, why they alone are able to reproduce life through their children, and why they are able to tend and nurture these children in their houses. It is also why women alone manage to grow crops of manioc tubers in their gardens. These tubers are their other children.

Today, the music of the Juruparí instruments is the living soul of the members of the clan and of its ancestors. It also guarantees the fertility and reproduction of the human and natural worlds. To keep women in their place and to stop things from going wrong again, the men were told to seek a 'perfect woman', one who was obedient and sexually chaste, who did not gossip and who did not pry into men's secrets – in other words, a woman disciplined and continent in all respects and at every aperture of her body. From then on, whilst women were made to become the obligatory audience for the men's ritual music, they were forbidden to see the flutes and trumpets that makes this music. Above all else, they were forbidden to play these or any other kind of wind instrument or to wear any form of feather ornament. Such restrictions apply throughout indigenous Amazonia and, in the case of the Tukano that we are concerned with here, women are even prohibited from whistling.

Galen in Amazonia

We can now begin to put these pieces together to see how the human body, as it appears in an Amazonian story about unruly women who steal flutes, has echoes in Gordon's (2004) account of the unruly singers of Monteverdi's Italy.

Let us return to the three principles which I have used to summarize some of Galen's anatomical ideas, starting with the horizontal analogy between penis and vagina. The stories summarized above describe an initial state where there are no sexual organs and no sex. Our story begins with a deity who acts as a man by using her mouth and mouth-associated objects of men's ritual to reproduce - she fertilizes coca in a gourd by blowing smoke from a cigar held in a cigar holder. But this deity would also be female insofar as she is made pregnant by eating ash from a cigar or coca or tapioca from a gourd, an inference reinforced when it is remembered that her own sexual organs derive from a cigar holder (see above). The pairing of cigar and the leg-like prongs of the cigar holder requires no further comment.

This insemination by cigar is but one of several incidents in NW Amazonian mythology where blowing smoke or music-laden breath from the mouth figures as insemination, where women become pregnant by ingesting substances through the mouth, and where vomiting and regurgitating from the mouth figure as a mode of creation and child-birth. We saw how Juruparí restores life, and effectively gives birth, to the dead initiate boys who he has previously ingested by vomiting them into an enclosure from which they then emerge as initiated young men. In an origin story common to all the Tukanoan-speakers, a deity creates the first ancestors in spirit form by vomiting them from his mouth as the feather head-dresses, cigar-holders and other paraphernalia of Tukano ritual. The voices of these spirit ancestors are the sounds of the flutes and trumpets and their dazzling appearance is that of the coloured feather ornaments that adorn not only the heads of the men who play flutes but also the flutes themselves. These disembodied spirit ancestors, still in the form of coloured feather ornaments, then enter the body of an anaconda-canoe who swims upstream with them to the centre of the world, the equatorial region where they now live. The journey inside this snake-canoe is likened to a gestation and, on arrival, the anaconda vomits up the spirit-ancestors to be re-born in the embodied form of true human beings.

Our story then tells us that, when the men re-gained control of the flutes stolen by the women, the women kept one of these flutes inside their bodies as their own vaginas and that it was they, rather than the men, who now began to menstruate. Other versions of the story have it that it was the men who created the women's genitals and caused then to menstruate by using their flutes to rape the women as a punishment for stepping out of line.

Either way, this incident suggests that women's genitals are interior, inverted analogues of men's mouth-blown flutes and that the music that flows from these flutes is the analogue of a menstrual flow. The same incident would also suggest that the flutes and trumpets to which a young boy first gains access when he undergoes initiation are the counterparts of the boy's own developing genitals as he reaches sexual maturity, a suggestion reinforced by the fact that, during initiation, the Juruparí instruments are played directly over the genitals of the initiate boys. This would imply that the music that comes from these instruments is akin to semen. Thus, through the mediation of musical instruments, in Tukanoan mythology and ritual, we find the horizontal analogy between male and female genitals that we have already noted in Galen's anatomy. But, for both young men and young women, the opening of the body that occurs at puberty must be put firmly under social control. If the story of Juruparí suggests that excessive closure of the body's apertures is tantamount to physical and social sterility, it also teaches that excessive openness threatens chaos, destruction and death.

The associations between Juruparí flutes and the male and female genitals suggests that life-giving flute music is menstrual blood in a seminal, male mode or, conversely, that menstrual blood is seminal flute music in a female mode. We have already met this equivalence between music, blood and semen in the case of women singers in Monteverdi's Italy. The same equivalence is also suggested in the two-way switch between flutes and menstruation that occurs in the Juruparí story and yet again in the belief that when initiate boys are first shown the Juruparí instruments, they are in a state equivalent to menstruation.

This mythological emphasis on reproduction via the mouth is consistent with our second principle, that of a vertical analogy between the upper and lower body or between genitals and mouth, an analogy that Gordon (2004) detects in the correspondence between a renaissance singer's open mouth and her genitals below and in the evocation of blood and semen in her song-laden breath. But instead of a female mouth producing provocative song in the ritualized context of the theatre, the mythology we are examining recasts this as a vertical contrast between two different, gendered modes of reproduction. On the one hand, we have the initial state of mythological time in which reproduction involves androgynous ancestral beings with no sexual organs who reproduce through their mouths, either by vomiting up feather ornaments or by blowing out smoke inhaled from cigars or by blowing breath through musical instruments to produce music – the similar construction of cigars and trumpets, the former made from spiralled leaf and the latter from spiralled bark, is striking. Like semen that flows up the penis and vagina, breath, smoke and vomit all flow upwards through the throat and against the current of the digestive tube, the same upstream direction taken by the anaconda-canoe on an ancestral journey that converts spirits in the form of feather ornaments and other ritual goods into true human beings. The sexual connotations of these asexual, ritual modes of reproduction where substances spew upwards from the mouth are obvious enough.

On the other hand, this pre-genital, asexual and upper-body mode of reproduction contrasts with the lower-body genital and sexual reproduction of the normal human beings of post-mythological times. This contrast is made into one of gender because the asexual reproduction of mythological times is also what takes place in the context of male ritual activities and because these activities are opposed to female menstruation and childbirth, two processes that both involve a downward flow through the body that runs in the same downstream direction as the other natural bodily processes of eating, digestion and excretion.

In mythology, the contrast between the flutes of men's ritual reproduction and the menstruation of female sexual reproduction becomes a switch between two alternatives – what one sex gains the other must lose. In the rituals where the flutes and trumpets are used, this contrast is given dramatic expression when the women are first allowed back into the house at the end of the rite. To avoid any possible contamination from female blood, before any women is allowed to re-enter, the men ostentatiously remove all ritual goods and go to sit briefly with them outside.

The effect of this contrast is to down-play or obscure the male role in sexual reproduction and to displace it to the superior realm of ritual reproduction that is held to ensure not just the human fertility but also the reproduction of all forms of life in the gardens, forest and rivers. In this ritual mode of reproduction, fertile, semen-like breath is made visible and audible as smoke that is inhaled from a cigar and then blown noisily from the mouth. This same breath acquires yet more sound and substance when it is exhaled through a flute to produce music.

The visible counterparts of this music are the ornaments that the women stole from the men along with their flutes, the same ornaments that were vomited up by the deity, the vehicles of soul that later gave rise to human beings. Tukano consider hair and feathers to be the same kind of material and refer to them using the same word. Feather ornaments are worn by men whenever they dance and appear to sprout from their heads while breath flows from their mouths as song and music. These feather ornaments are a form of wig that enhances the volume and visual impact of the hair on the head (Figure 6). Furthermore, along with tobacco smoke, feathers, especially in the form of down or fine filaments, are another material that makes air in motion visible.[11]

We have already encountered this same pairing between music and hair in the figure of Jurupari, a being whose colourful, snake-like body is full of holes that emit a stream of hair and sound. And we meet this pairing again in the belief that a woman's menstrual period begins when the hair that she normally wears behind her head falls in front of her eyes – yet one more example of the very common association between the flow of hair from the head and the flow of menstrual blood explored by Karadimas (2010).

We can now begin to see that, in the context of mythology and ritual, hair, feathers, blood, semen and breath charged with smoke or music all appear as closely-allied substances that flow from various apertures in the body. These substances are all manifestations of *üsü* or *ehêri põ'ra*, terms that refer to 'breath', 'heart', 'vital force', 'soul', 'mind', 'feeling', 'intuition' and 'life force'.[12] This is close to the *pneuma* of the medical writers and philosophers of antiquity, a term which also brings together air, breath, spirit, soul as a material substance contained in semen (Wikipedia 2019). This brings me back to the equivalence or fungibility of different body substances, the third of three principles that I have used to summarize some of Galen's anatomical and physiological ideas – and here I am especially interested in the equivalence

---

[11]This might explain why shamans officiating at major rituals wear dollops of white down on their heads. The shaman's principal activity is to blow protective spells into the coca, cigars, manioc beer and *yage/ayahuasca* consumed during the rite so the down might suggest his superabundant breath.

[12]The Barasana term *üsü* has connotations of heart, lungs, breathing and vitality and of that which generates life and animates persons or social groups. *Üsü* is also manifest in the personal names and feather ornaments owned by particular Tukanoan groups. With names passing between alternating generations and ornaments inherited from the past, buried with the dead and returning to animate future generations, names and ornaments are linked as enduring and circulating sources of vitality and identity (see Hugh-Jones 2002, 2006, 2018).

Considerations such as these explain why the Tukano origin story concerns a journey of feather ornaments giving rise to human beings, why the journey takes 9-months, and why shamans recapitulate this origin story during the shamanic procedures involved in name-giving (see Hugh-Jones 2002, 2006) and why Barasana involved in Bible translation use *üsü* to render 'soul/*alma*' (see Jones and Jones 2009, 319). Much the same considerations would apply to the Tukano *êheri põ'ra* (see Ramirez 2019, 47).

Figure 6. Barasana dancer wearing feather ornaments (Photo Brian Moser).

between a woman's blood and her song-laden breath that we have encountered earlier.

This equivalence between blood, music and musical ritual more generally is made quite explicit in the story of *Kahi* or *Kaapi*, the name and personification of the hallucinogenic ayahuasca or yagé (Banisteriopsis) that Tukano use in their rituals. One of the effects of this drug is to produce intense sensations of synaesthesia so that one sees music as visual patterns and hears colours and designs as music. The story of *Kahi* is a close variant of the story of Jurupari, the two sometimes appearing as one and the same person and sometimes as twin brothers of twin sisters. The circumstances of their births from mothers who initially lack the necessary aperture are also more or less identical. In addition, just as the men first take Jurupari from his mother and later dismember his bones to make flutes and trumpets, so also do they take the new-born *Kahi* from his mother and dismember his body in the form of the Banisteriopsis vines that they now use to prepare yagé as a drink. But the story also tells us that it was precisely the blood that flowed from *Kahi*'s mother's body at the moment of his birth that gave rise not only to the synaesthetic hallucinations that one experiences when one drinks ayahuasca but also to all species of coloured and patterned snakes, to all the different coloured birds whose feathers are used to make ornaments, to all forms of song and music and to all forms of pattern or design such as those that are painted on the body and on stools, woven into basketry, or painted on the outside of houses.

True to the philosophical bent that Lévi-Strauss detected in Tukano mythology, here in the story of *Kahi*, we find an explicit theory of the synaesthetic effects of ayahuasca that locates these effects in a flow of blood and which presents blood, feathers, ornaments, music and song as all transformations of the same kind of stuff. This same synaesthesia is also apparent in the music and hair that flows from Jurupari's body. Finally, the same story also recounts that it was only after Banisteriopsis vines had been created and distributed amongst human groups that these different groups began to intermarry, that the bodies of women and men became different from one another, and that sexual relations, normal pregnancy and genital birth took their place alongside men's ritual.

Conclusion

I began this chapter by suggesting that the Amerindian philosophy of body tubes and apertures that Lévi-Strauss detects in Amerindian mythology about pottery is condensed more succinctly in the story of Jurupari – and more succinctly still in the Barasana aphorism that 'the body is a flute'. Ironically, and as noted previously, it was precisely its explicit, reflective philosophical bent that led Lévi-Strauss to largely exclude NW Amazonian mythology from the first four volumes of his otherwise comprehensive *Mythologiques*. Lévi-Strauss seems later to have come to recognize this irony. Apropos of the *Jealous Potter*, a book that is, in effect *Mythologiques* vol. 5, he writes somewhat ruefully ' a complex argument, spread over one hundred pages had been necessary to demonstrate that myths about the origins of pottery and those about dwarfs without anuses belong to the same set', something that is already quite explicit in a single Tatuyo myth (2013, 247 my trans).

Like the mythology discussed in *The Jealous Potter*, the story of Jurupari is structured around the absence or presence of body apertures, their open or closed state, and their proper control as indices of the moral state of the persons involved. This proper control implies the balance between two extremes on which all life depends. At one extreme is sterile closure represented by a Jurupari with no mouth and his mother with no genitals. At the other extreme is excessive, life-threatening openness of a Jurupari as cannibal monster and of his mother as sexually devouring ogress.

But this balance is also internal to the paradoxical figure of Jurupari himself, for Jurupari is at once an uptight, hairy sloth and a devouring monster with gaping mouth and wide-open anus. This same balance can also be seen in the striking contrast between two indigenous portrayals of Jurupari, one as a monstrous wild animal with the hands and feet of a sloth, the other as a beautiful human dancer with an ornamented body and painted face (Figure 3, Figure 4). Here, instead of hair, it is music that emerges from the holes in Jurupari's body, music that is rendered in visually as another form of ornament. Finally, this balance can also be heard in the contrast between the two kinds of musical instruments that represent Jurupari on earth today, a contrast between the raucous, threatening sound of trumpets – Juruparí in his loud, howler monkey mode – and the sweeter, restrained sound of flutes – Juruparí in his quieter, sloth-like mode.

I have also suggested that the body that lies behind this Tukano version of a philosophy of body-tubes may not be the same official, modern body we take for granted and that we may gain insight into this different body by following up parallels between Tukano mythology, the medical writings of Galen and ideas about female singers in Monteverdi's Italy. Some may find this Amazonian ethnography somewhat outlandish and these classical and renaissance parallels a bit far-fetched. I would respond that these Amazonian ideas and their European parallels are not so exotic as they might seem at first sight. This is because it is part of daily experience that the body as a whole is tubular in structure, that it is made up of various other tubes and that the whole of life depends on various flows in and out of these tubes. This shared experience is the basis of what we might call a common-sense or folk anatomy and physiology common to most people in the world. It is also a rich source of metaphors that play upon analogies between body tubes and the objects we make and use.

In this very general sense, Lévi-Strauss' philosophy of body tubes is already quite familiar to us. The ideas behind this philosophy are also familiar to us in the surrealist transpositions between the top and bottom halves of the body in Magritte's painting 'Le Viol' and in the long, phallic noses of carnival masks and New Guinea sculpture. They are also implicit in the colourful but unofficial words and idioms of French and English slang, some of which date back to mediaeval times. And here too we find musical instruments standing for body parts, bodily functions and aspects of character and behaviour. My Tukano friends would be amused to learn that 'la clarinette' figures as a French slang expression for 'penis'.

But today, in the modern West, this way of thinking has been overlain and driven underground by what Elias (1969) called the civilizing process. In this process, the body of folk anatomy and physiology has been obscured by a modern, scientific body that is married to a religious emphasis on thought, speech, decorum and the soul which eschews bodily functions and sexual behaviour. The result is that the mythology of NW Amazonia seems to belong more to Rabelais' Renaissance world of (2006) *Gargantua and Pantagruel* and to the grotesque body of Bahktin's (1941) carnivalesque than it does to the polite society of today. But is also for reasons such as these that this same mythology has relevance beyond the rather specialized field of Amazonian ethnography for it also speaks of matters that are close to home.[13]

---

[13]It is interesting to note here that many explorers and missionaries from the eighteenth to nineteenth centuries characterized the sound of Juruparí flutes as melodious and the sound of the trumpets as wild and devilish. With Juruparí trumpets amplifying a 'raspberry' blown with the lips and with Juruparí's mouth or anus as a cave, there is a degree of overlap here with European ideas of the grotesque, a term itself derived from 'grotto' or cave (see also Tradii n.d.).

In late Renaissance Italy, where a woman's mouth had sexual connotations and where singing was seen as an essentially male activity associated with masculine power and dominance, women singers had to tread a fine line between welcome titillation and unacceptable transgression. In the upper Rio Negro region, as elsewhere in Amazonia, women are rarely if ever allowed to play any kind of wind instrument and almost never wear ornaments made from feathers. However, whereas Renaissance men doubtless found a singer's transgression across bodily and social boundaries both threatening and titillating, Amazonian men find the thought of a woman wearing feather ornaments, seeing flutes or playing any kind of wind instrument as well beyond the pale. I suggest that ideas from Renaissance Italy, transposed to an Amazonian context, may help us in understanding why this should be so. Amazonian men find the thought of a woman wearing feather ornaments, seeing flutes or playing any kind of wind instrument as something beyond the pale of normal life, relegating it to the more virtual world of myth and ritual.[14] I suggest that ideas from Renaissance Italy, transposed to an Amazonian context, may help us in understanding why this should be so.

Finally, the idea that the body is not given by nature but is rather something that must be constructed through prolonged and careful social and ritual intervention has played a strategic role in our understanding of the peoples and cultures of Amazonia. I also suggest that the ideas I have discussed may shed new light on the anatomy and physiology of this constructed Amazonian body. But here I would introduce a word of caution. As mentioned earlier, in her critique of Laqueur's (1990) thesis concerning the transition from a one-sex to two-sex, two genders model of the body as marking the transition from the mediaeval era to modernity, King (2013) provides evidence for the co-existence of various alternative ancient Greek views on female anatomy and physiology and on the disputed issue of the female contribution to conception.

This should alert us to the possibility of a similar plurality of views on issues to do with anatomy, physiology and the mechanisms of conception in the context of Tukano culture. This is rendered more likely not only because these views concern matters many of which are hidden from view but also because they lie in an area where knowledge derived from observation intersects with ideology and in a cultural context where gender and hierarchy are especially salient.[15] For these reasons, I would not wish to suggest that the body that we meet in Tukano mythology and ritual is the one and only body in all of Tukano thought. By the same token, the body that we find in the biblical stories of immaculate conception and virgin birth will tell us little about ordinary people's common-sense understandings of reproduction.

---

[14]See, for e.g. Mello and Ignez (2011), Fausto et al. (2011).
[15]See Hugh-Jones (2001).

Anthropologists have often interpreted the story and rituals of Jurupari as serving to reinforce men's dominance over women. But the stories of Jurupari and *Kahi* might also be read as stories about two gendered interpretations of the reproduction of life, one located most obviously in women's bodies, the other in the noisy, showy paraphernalia of men's ritual. These two interpretations of reproduction are fused in the androgynous but still female Barasana deity Romi Kumu and her equivalents in the mythology of other Tukano groups. Romi translates as 'woman' whilst the role of kumu ('shaman, ritual expert') is normally associated with men and carries masculine connotations.

Disclosure statement

No potential conflict of interest was reported by the author(s).

References

Armstrong, David, and Anne Ellis Hanson. 1986. "Two Notes on Greek Tragedy." *Bulletin of the Institute of Classical Studies* 33: 97–100.

Bahktin, Mikhail. 1941. *Rabelais and his World*. Bloomington: Indiana University Press.

Barcelos Neto, Aristoteles. 2016. "The Wauja Snake-Basket: Myth and the Conceptual Imagination of Material Culture in Amazonia." *Mundo Amazónico* 7 (1-2): 115–136.

Elias, Norbert. 1969. *The Civilizing Process, Vol. I. The History of Manners*. Oxford: Blackwell.

Fausto, Carlos, Leonardo Sette, and Takumã Kuikuro. 2011. *As Hiper-Mulheres*. Feature Film (80'). Associação Indígena Kuikuro do Alto Xingu, Documenta Kuikuro – Museu Nacional and Vídeo nas Aldeias.

Gordon, Bonnie. 2004. *Monteverdi's Unruly Women*. Cambridge: Cambridge University Press.

Hugh-Jones, Stephen. 1979. *The Palm and the Pleiades*. Cambridge: Cambridge University Press.

Hugh-Jones, Stephen. 2001. "The Gender of Some Amazonian Gifts; an Experiment with an Experiment." In *Gender in Amazonia and Melanesia*, edited by Thomas Gregor and Donald Tuzin, 245–278. Berkeley: University of California Press.

Hugh-Jones, Stephen. 2002. "Nomes secretos e riqueza visível: nominação no noroeste amazônico." *Mana* 8 (2): 45–68.

Hugh-Jones, Stephen. 2006. "The Substance of Northwest Amazonian Names." In *The Anthropology of Names and Naming*, edited by Gabrielle vom Bruck and Barbara Bodenhorn, 74–96. Cambridge: Cambridge University Press.

Hugh-Jones, Stephen. 2018. "Su riqueza es nuestra riqueza: perspectivas interculturales de objetos o *gaheuni*." In *Objetos como testigos del contacto cultural. Perspectivas interculturales de la historia y del presente de las poblaciones indígenas del alto río Negro (Brasil/Colombia)*, edited by Michael Kraus, Ernst Halbmayer, and Ingrid Kummels, 197–226. Berlin: Ibero-Amerikaniches Institut/Gebr. Mann Verlag.

Jones, Wendell, and Paula Jones. 2009. *Diccionario Bilingüe Eduria & Barasana – Español, Español – Eduria & Barasana*. Bogotá: Editorial Fundación para el Desarrollo de los Pueblos Marginados.

Karadimas, Dimitri. 2010. "Pilosité et sang: une problématique." In *Poils et sang*, edited by Dimitri Karadimas, 13–26. Paris: L'Herne.

King, Helen. 2013. *The One Sex Body on Trial*. London: Routledge.

Koch-Grünberg, Theodor. 1967. *Zwei Jahre Unter den Indianern*. Graz: Akademische Druck.

LaMay, Thomasin. 2007. "Review of Monteverdi's Unruly Women: The Power of Song in Early Modern Italy." *Women and Music* 11: 101–106.

Lana, Luiz (Tõrãmú Kehíri), and Firmiano Lana (Umúsin Pãrõkumu). 1995. *Antes o Mundo Não Existia*. São João Batista do Rio Tiquié: UNIRT/ São Gabriel da Cachoeira: FOIRN.

Laqueur, Thomas. 1990. *Making Sex*. Cambridge, MA: Harvard University Press.

Lévi-Strauss, Claude. 1970. *The Raw and the Cooked, tr. John and Doreen Weightman*. London: Jonathan Cape.

Lévi-Strauss, Claude. 1973. *From Honey to Ashes, tr. John and Doreen Weightman*. London: Jonathan Cape.

Lévi-Strauss, Claude. 1984. *Paroles données*. Paris: Plon.

Lévi-Strauss, Claude. 1987. *Anthropology and Myth: Lectures 1951–1982*. Oxford: Blackwell.

Lévi-Strauss, Claude. 1988. *The Jealous Potter tr. Bénédicte Chorier*. Chicago: University of Chicago Press.

Lévi-Strauss, Claude. 2013. 'La preuve par mythe neuf', dans *Nous sommes tous des cannibales*. Paris: Seuil. 243–251.

Meira, Márcio. 2017. *A persistência do aviamento: colonialismo e história indígena no noroeste Amazônico*. PhD Thesis, Federal University of Rio de Janeiro.

Mello, Cruz, and Maria Ignez. 2011. "The Ritual of Yamariukuma and the Kawoká Flutes." In *Burst of Breath. Indigenous Wind Instruments in Lowland South America*, edited by Jonathan Hill and Jean-Pierre Chaumeil, 257–276. Lincoln & London: University of Nebraska Press.

Oliveira, Tiago. 2015. "Os baniwa, os artefactos e a cultura materia no alto rio negro." Ph.D. Thesis. Universidade Federale do Rio de Janeiro, Museu Nacional.

Preuss, Anthony. 1997. "Galen's Criticism of Aristotle's Conception Theory." *Journal of the History of Biology* 10 (1): 65–85.

Rabelais, François. 2006. *Gargantua and Pantagruel*. London: Penguin.

Ramirez, Henri. 2019. "A Fala Tukano dos Ye'pâ-Masa". Tomo II Diccionário Manaus: Inspetoria Salesiana Missionária da Amazônia CEDEM.

Straddelli, Ermanno. 1929. "Vocabularios da lingua geral portuguez-nheêngatú e nheêngatú-portuguez." *Revista do Instituto Historico e Geographico Brasileiro, Tomo 104* 158: 9–768.

Tradii, Laura. n.d. "The grotesque: mischief and wonder in renaissance art". http://www.dilettantearmy.com/articles/grotesque. Consulted 19.04.2020.

Van Velthem, Lucia. 2003. *O belo é a fera: a estética da produção e da predação entre os Wayana*. Lisboa: Assírio & Alvim/Museu Nacional de Etnologia.

Vianna, João. 2017. *Kowai e os nacidos*. PhD Thesis, Universidade Federal de Santa Catarina.

Wikipedia. 2019. "Pneuma – Wikipedia". https://en.wikipedia.org/wiki/Pneuma. Accessed 03.12.2019.

Wright, Robin. 2018. "The Kuwai Religions of Northern Arawak-Speaking Peoples: Initiation, Shamanism, and Nature Religions of the Amazon and Orinoco." *Boletín de Antropología* 33 (55): 123–150.

Index

Note: Figures are indicated by *italics*. Endnotes are indicated by the page number followed by 'n' and the endnote number e.g., 20n1 refers to endnote 1 on page 20.

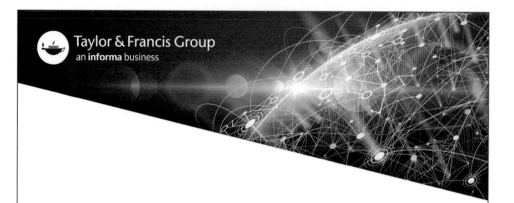